EARTH SCIENCES IN THE 21ST CENTURY

RED SEA

HISTORICAL SIGNIFICANCE, PROPERTIES AND ECONOMIC IMPORTANCE

Earth Sciences in the 21st Century

Additional books in this series can be found on Nova's website under the Series tab.

Additional e-books in this series can be found on Nova's website under the eBooks tab.

EARTH SCIENCES IN THE 21ST CENTURY

RED SEA

HISTORICAL SIGNIFICANCE, PROPERTIES AND ECONOMIC IMPORTANCE

DELBERT ROBBINS
EDITOR

nova
science publishers
New York

Library of Congress Cataloging-in-Publication Data

ISBN: 978-1-53613-200-7

Published by Nova Science Publishers, Inc. † New York

CONTENTS

PREFACE

Red Sea: Historical Significance, Properties and Economic Importance opens with a discussion on oceanic spreading beginning in the Pliocene (~5 Ma) (after advanced continental drifting) and continuing to the present day, as demonstrated by the intense volcanic and seismic activity in the area during the last 10,000 years. The authors present geochemical features that are typical of mid ocean ridge basalts (MORB), which allow the Red Sea to be regarded as a newborn ocean. Additionally, its development resembles the early stages of the Atlantic Ocean. Next, this compilation discusses the Wadi Masilah Basin in the southeastern most part of Yemen, which hosts promising base metal and barite prospects. This terrain has a complex geological and structural history, including oblique divergence (transtension) related to the opening phases of the greater Red Sea. The fisheries reserve of the Sudanese Red Sea is dominated by different assemblages of ecologically and commercially important stock of invertebrates, demersal finfish and pelagic fish. The authors address how the national fisheries standing stock remains largely under-exploited with the exception of Mother of Pearl (*Pinctada margeritifera*) and Trochus (*Tectus dentatus*) which are efficiently utilized.

Chapter 1 - During much of the Phanerozoic until the Oligocene, the African and Arabian plates were part of a single shield. The history of its dislocation via the Red Sea opening starts in the late Oligocene (~24 Ma), when the Arabian plate separated from that of Nubia through a rupture of the Precambrian shield and rotated counter-clockwise. After advanced continental rifting, oceanic spreading began in the Pliocene (~5 Ma) and continues to the present day, as demonstrated by the intense volcanic and seismic activity in the area during the last 10,000 years. Seafloor magnetic anomalies and normal faulting earthquakes, typically associated with spreading ridges, occur in the Red Sea south of 20°N, where a deep axial trough is present. To the north, only few isolated anomalies are identified, and the sea is floored by rotated fault blocks. South of 16°N the axial trough narrows and some groups of islands emerge, the last of which was formed in 2013 in the Zubair Archipelago. The northern area is interpreted as continental crust being stretched and faulted with only minor volcanic activity, whereas the southern area has evolved to the point that new lithosphere is being produced at a spreading axis. This situation has been viewed as transient until the extension in the northern area evolves into seafloor spreading. The southern deep axial trough is the focus of intense submarine volcanic activity. The geochemistry of erupted magmas conveys valuable information about the underlying mantle. They are low potassium tholeiitic basalts with K2O $< 0.2\%$ and TiO2 $< 2.0\%$. Their Sr and Nd isotope ratios are ≈ 0.703 and ≈ 0.513, respectively. These geochemical features are typical of mid ocean ridge basalts (MORB), and allow the Red Sea to be regarded as a newborn ocean, whose development likely resembles that of the Atlantic Ocean in its early stages, and which may one day become the sixth ocean of the planet. It is interesting to note that the erupted lavas are characterized by geochemical variations along the Red Sea. The basalts of the central part exhibit geochemical parameters that resemble those of MORB. Both north and south, the basalts display higher contents of incompatible elements, higher Sr and lower Nd isotope ratios, as well as higher Pb isotopes ($^{206}Pb/^{204}Pb$ up to 19.608), than those recorded in the central part, suggesting a contribution by "enriched" mantle components.

Chapter 2 - Yemen is characterized by a considerable number of metallic and non-metallic raw materials, including base metals, precious metals, and dimension stones. The Wadi Masilah Basin, in the southeasternmost part of Yemen, hosts very promising base metal and barite prospects. This terrain has undergone a complex geological and structural history, including oblique divergence (transtension) related to the opening phases of the greater Red Sea – Gulf of Aden rift system, with NW-SE and NE-SW major fault zones. Barite occurs either singly or in association with galena, hemimorphite, willemite, zincite, smithsonite, descloizite, celestine, quartz, calcite, dolomite, iron and manganese oxides, and gypsum, connected mainly to the Cretaceous carbonates (Saa'r Formation). The mineralisations crop out as either vein fillings or stratiform type. $BaSO_4$ contents between 95 and 98% and specific gravity values up to 4.37 g/cm^3 are indicative of high quality ores. Sulfide mineralisations consist of sphalerite, galena and pyrite. They occur in association with barite veins or form fracture-filling in limestones and silicified dolostones. Available data indicate several lead-zinc ore bodies at various depths, some of which are considerably large with dimensions attaining 60 m x 30 m x 50 m, and tonnages reaching thousands of tons. Nonsulfide minerals comprise hemimorphite, willemite, smithsonite, hydrozincite, and zincite, as well as cerussite, hydrocerussite, anglesite, wulfenite, and rarely pyromorphite. Zinc contents in the mineralised veins range from low (1-8 wt.%) to high grades (30-50 wt.%). The above mentioned minerals occur in a variety of structures, such as: shear zones, fractures, vugs, cavities, brecciated veins, and open spaces of the vein barite sidewalls, and connected mainly to the dolostone sequence. Vanadium ores mostly occur as descloizite and rarely vanadinite, in association with barite, calcitized veins and brecciated zones, and dolomitic limestones, with vanadium contents up to 12.30 wt.%. These prospects contain one of the most remarkable occurrences of descloizite known, and besides lead-zinc, descloizite is probably the mineral resource for which the Wadi Masilah Basin is most promising. Mn and Fe ores are also present, whose mineralogy consists of hollandite, romanechite, cryptomelane, pyrolusite, hematite, goethite and limonite. They include several discordant irregular veins. The results of geochemical analyses reveal high quality Mn ores, with an average MnO content of 64%. The orebodies in all prospects consist mainly of fracture-filling in limestones, dolostones, and silicified limestones. They originated from both hydrothermal processes and the action of slightly acidic groundwater. The prospecting area is structurally controlled and located in a NW-SE major fault zone and its parallel lineaments. Tectonic uplift and intensive fracturing systems, along with favorable climatic conditions, gave rise to the development of supergene Zn, Pb, V, Mn and Fe mineralisations.

Chapter 3 - The Sudanese Red Sea is considered to be unique in many aspects. It is still pristine with rich and diversified ecosystems and high biodiversity in addition to other potential resources of high economic value. Among these resources is a rich fisheries reserve dominated by different assemblages of an ecologically and commercially important stock of invertebrates, demersal finfish and pelagic fish. The national fisheries standing stock, excluding the shared stock, remains largely under-exploited with the exception of Mother of Pearl (*Pinctada margeritifera*) and Trochus (*Tectus dentatus*) which are efficiently utilized. The fisheries resource if fully exploited will greatly contribute to the food security of the region and provide sources of income to the local communities. Official statements and reviews including international and regional reports by the Food and Agriculture Organization of the United Nations (FAO), United Kingdom Overseas Development Administration (ODA), Regional Organization for the Conservation of the Environment of the Red Sea and Gulf of Aden (PERSEGA) and Sudan National Report have identified a number of challenges that the Sudanese Marine Fisheries Sector is facing in order to increase the catch yield.

In: Red Sea
Editor: Delbert Robbins
ISBN: 978-1-53613-200-7

Chapter 1

THE RED SEA: A NEWBORN OCEAN

Laura Pinarelli, PhD*
Institute of Geosciences and Earth Resources, Section of Florence,
National Research Council (CNR), Florence, Italy

ABSTRACT

During much of the Phanerozoic until the Oligocene, the African and Arabian plates were part of a single shield. The history of its dislocation via the Red Sea opening starts in the late Oligocene (~24 Ma), when the Arabian plate separated from that of Nubia through a rupture of the Precambrian shield and rotated counter-clockwise. After advanced continental rifting, oceanic spreading began in the Pliocene (~5 Ma) and continues to the present day, as demonstrated by the intense volcanic and seismic activity in the area during the last 10,000 years.

Seafloor magnetic anomalies and normal faulting earthquakes, typically associated with spreading ridges, occur in the Red Sea south of 20°N, where a deep axial trough is present. To the north, only few isolated anomalies are identified, and the sea is floored by rotated fault blocks. South of 16°N the axial trough narrows and some groups of islands emerge, the last of which was formed in 2013 in the Zubair Archipelago. The northern area is interpreted as continental crust being

* Corresponding Author Email: lapina@igg.cnr.it.

stretched and faulted with only minor volcanic activity, whereas the southern area has evolved to the point that new lithosphere is being produced at a spreading axis. This situation has been viewed as transient until the extension in the northern area evolves into seafloor spreading.

The southern deep axial trough is the focus of intense submarine volcanic activity. The geochemistry of erupted magmas conveys valuable information about the underlying mantle. They are low potassium tholeiitic basalts with K2O < 0.2% and TiO2 < 2.0%. Their Sr and Nd isotope ratios are ≈ 0.703 and ≈ 0.513, respectively. These geochemical features are typical of mid ocean ridge basalts (MORB), and allow the Red Sea to be regarded as a newborn ocean, whose development likely resembles that of the Atlantic Ocean in its early stages, and which may one day become the sixth ocean of the planet. It is interesting to note that the erupted lavas are characterized by geochemical variations along the Red Sea. The basalts of the central part exhibit geochemical parameters that resemble those of MORB. Both north and south, the basalts display higher contents of incompatible elements, higher Sr and lower Nd isotope ratios, as well as higher Pb isotopes ($^{206}Pb/^{204}Pb$ up to 19.608), than those recorded in the central part, suggesting a contribution by "enriched" mantle components.

Keywords: Red Sea, basalts, geochemistry, isotopes

1. Introduction

Since the 1960s the Plate Tectonic scientific revolution demonstrated that the Earth is subjected to cyclical processes, whereby the continents assemble in a single supercontinent, and then gradually fragment and disperse again. A fundamental step inside this cycle of continental breakup, dispersal, and reassembly, is the splitting of a continent and the growth of a new ocean. A major breakthrough in understanding the way the modern Earth works has been highlighted by the studies of the ocean floor: the Joint Oceanographic Institutions Deep Earth Sampling (JOIDES). This project began in 1969, continued with the Deep Sea Drilling Project (DSDP), and since 1976 it has been replaced by the International Phase of Ocean Drilling (IPOD) project. These projects have produced more than 500 boreholes in the seafloor worldwide, and the results achieved have been as outstanding as the plate-tectonic theory itself. They confirmed that the oceanic crust is everywhere younger than about 200 million years, and that the age of the oceanic crust calculated from the magnetic anomalies is close to the age of the overlying oceanic sediments estimated by micropaleontology.

Reconstructions of the geological past suggest that the continental masses from roughly 250 to 150 Ma were assembled in a single Supercontinent, called Pangea. The Red Sea is one of the components of a continent scale rift system located along a divergent plate boundary that, since the Oligocene, divide the African, Somalian, and Arabian plates (Figure 1), called Afro-Arabian Rift System (Baker 1970; Khan 1975). Besides the Red Sea, it includes the Gulf of Aden and the Ethiopian Rift, which together form a triple junction located at Afar (Gass 1970; McKenzie et al., 1970; Burke and Dewey 1974; Le Pichon and Francheteau 1978). Considering the evolutionary history of the planet, the Red Sea can be regarded as a fairly recent formation. It likely started forming $\approx$ 24 million years ago, when tectonic movements caused the separation of the African plate from the Eurasian one. The seismic activity that has always characterized the floor of this sea, causing an expansion of the basin between 10 and 16 mm/yr (Figure 1), is still present today.

2. Geological Setting

The Red Sea fills a long (about 2,250 km) and narrow (up to 350 km wide) depression that separates Africa from Arabia (Figure 2). Much of the Red Sea is only a few hundred meters deep, with the exception of the seafloor of a narrow valley that runs almost continuously along its axis, reaching more than 2,000 m in depth. The 130-year-old Suez Canal connects the Red Sea to the Mediterranean Sea, while the Bab el Mandab Strait divides the waters of the Red Sea from those of the Gulf of Aden–Indian Ocean.

During its geological history, the Red Sea basin has been separated several times from the ocean. The first temporary isolation dates back to about 5 million years ago, at the end of Tertiary, due to the rise of the backdrop of the Strait of Bab el Mandeb.

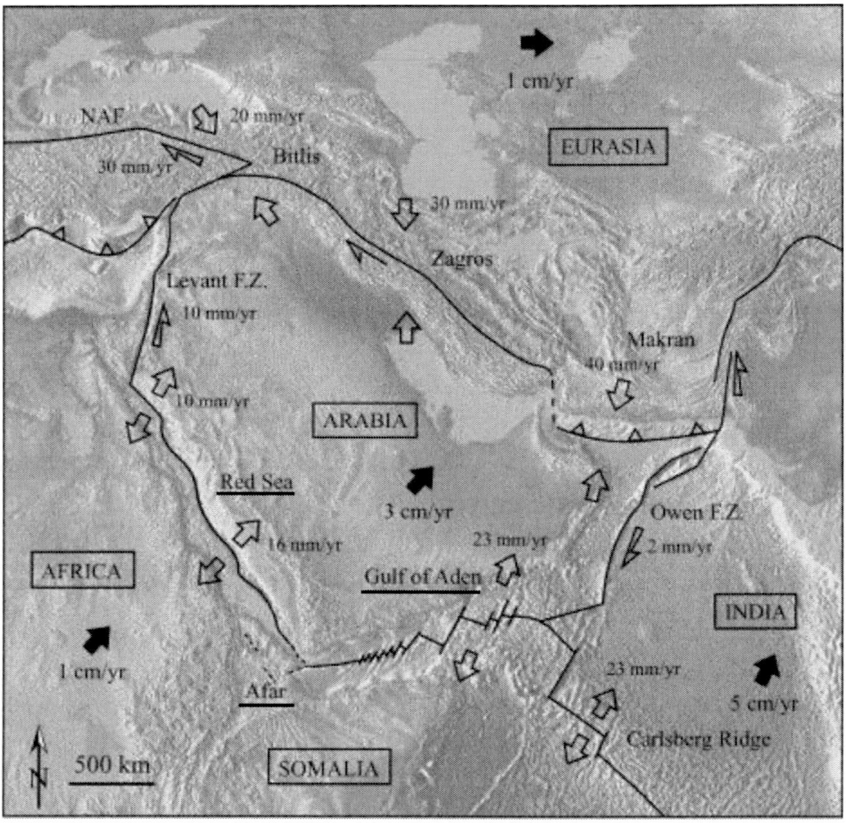

Figure 1. Tectonic framework of the Red Sea. Empty arrows mark direction and value of relative plate velocities. Black arrows represent absolute plate velocities (modified after Bellahsen et al., 2003).

The Red Sea is a classic example of a "closed sea." The limited supply of fresh water and the high evaporation rate have increased its salinity to 4.1%, against 3.5% of the Indian Ocean. Also the temperature (22°C in winter and 26°C in summer) is some degrees higher than that of the adjacent ocean. In addition, the area of the Bab el Mandeb Strait, 26 km wide and a maximum of 120 m deep, acts as an efficient isolation threshold with respect to the currents system of the Indian Ocean.

The average strike of the Red Sea from Suez to the Strait of Bab el Mandab is N30°W; however, the orientation of the Red Sea axial zone shows a few discontinuities, the most prominent of which is the Zabargad Fracture Zone (ZFZ in Figure 2), with stretches oriented ~5°–10°N (Crane and Bonatti 1987). This orientation is parallel to the Aqaba–Dead Sea Transform (ADST in Figure 2) that became active at roughly 14 Ma. As a consequence, the Red Sea modified its extension from rift normal to highly oblique, and the Gulf of Suez extension dramatically slowed although did not get to a complete stop (Bosworth et al., 2005).

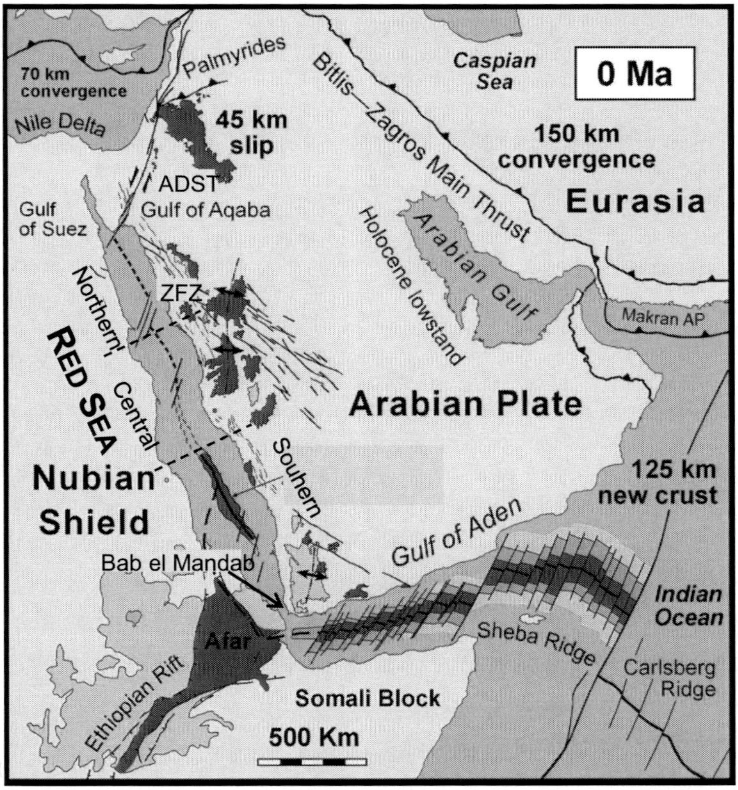

Figure 2. Schematic map of the Afro–Arabian rift system, including the Red Sea, the Ethiopian (East African) rift, Afar, and the gulfs of Aden, Aqaba and Suez (modified after Bosworth et al., 2005). ZFZ = Zabargad Fracture Zone. ADST = Aqaba–Dead Sea Transform. In dark grey is represented the Miocene-to-recent volcanism connected to the Red Sea rift.

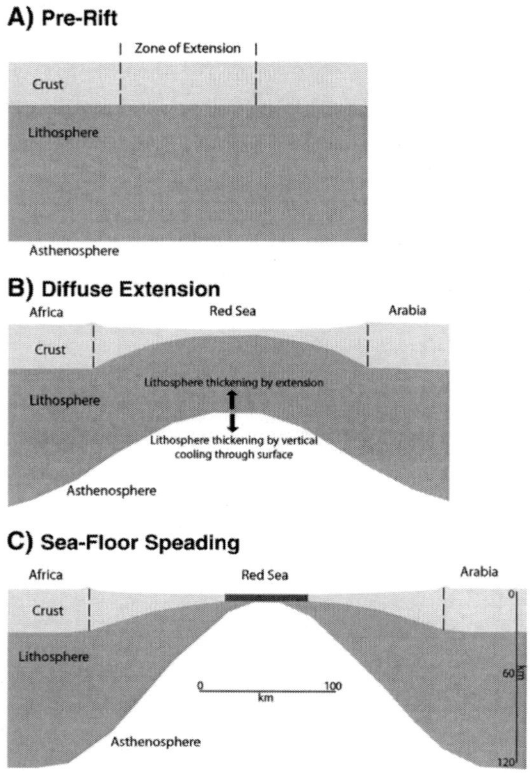

Figure 3. Development of the southern and central Red Sea continental margins and the beginning of the sea-floor spreading at 5 Ma (modified after Almalki et al., 2015).

According to most authors (e.g., Bosworth et al., 2005; Szymanski et al., 2016), continental rifting between Nubia and Arabia started in the late Oligocene (23-24 Ma) in both Northern and Southern parts of Red Sea, and it is still in progress. Plate tectonic reconstructions suggest that the oceanic crust started to be emplaced at ~5 Ma in the southern part of the Red Sea, whereas it has not started in its northern sector. Indeed, the southern Red Sea is presently an oceanic rift, whereas the northern end of the rift—the Gulf of Suez—is purely continental in character (Steckler 1985; Jarrige et al., 1986; Courtillot et al., 1987; Girdler and Southren 1987; Joffe and Garfunkel 1987; Moretti and Chénet 1987; McClusky et al., 2010; Keir et al., 2012; Mitchell and Park, 2014). The intervening areas — the central and the northern Red Sea — are transitional between aborted continental rifting, i.e., the Gulf of Suez, and well-defined oceanic rifting, i.e., the southern Red Sea. This led to extensional models in which different parts of the Red Sea (Northern, Central and Southern) were representing distinctive and different phases of the continental rifting with the development of a continental margin (Figure 3).

3. Magmatism in the Red Sea Inland

In the late Oligocene-early Miocene through to the present time, large volumes of flood basalts (Harrats) emplaced at discrete eruptive centres along the western margin of the Arabian plate, from the Gulf of Aden to the Mediterranean Sea (Figure 4). These plateau basalts are concentrated on the Arabian side of the Red Sea, without matching counterparts on the Nubian plate, and represent one of the largest areas of predominantly alkali-olivine basalts in the world, covering about 180,000 km^2. This magmatism is characterized by different chronological phases.

3.1. Older Harrats and Trap Basalts

Eruptions of the flood basalts began in Afar at 31 Ma, followed by rhyolitic eruptions ~1 Ma later, and then by a period of ~5 Ma of much more subdued outflowing of basalts and ignimbrites (Zumbo et al., 1995; Rochette et al., 1997; Chernet et al., 1998; George et al., 1998; Ukstins et al., 2002; Coulié et al., 2003). Smaller occurrences of rhyolites on the southern Sudanese margin (Kenea et al., 2001), and of basalts at the Older Harrat Hadan (Féraud et al., 1991; Sebai et al., 1991), show similar ages (30 Ma and 28–26 Ma, respectively). The Oligo-Miocene Yemen trap series (YTS, Mattash et al., 2013) have the same sequence as that observed at Afar, with basalts beginning at 31 Ma followed by repeated outflows from 30 to 16 Ma (Baker et al., 1994, 1996; Ukstins et al., 2002; Coulié et al., 2003).

In this period, intrusion of NW–SE striking basaltic dikes along the Yemen and Saudi Arabia Red Sea margins, Sinai, and northern Egypt, also occurred. This phase of magmatism was accompanied in the Red Sea by strong rift-normal extension and deposition of syn-tectonic sediments, followed shortly thereafter at ~20 Ma by the uplift of the western and southern margins of the Arabian Plate, creating what is now the Red Sea Escarpment and the mountains beyond.

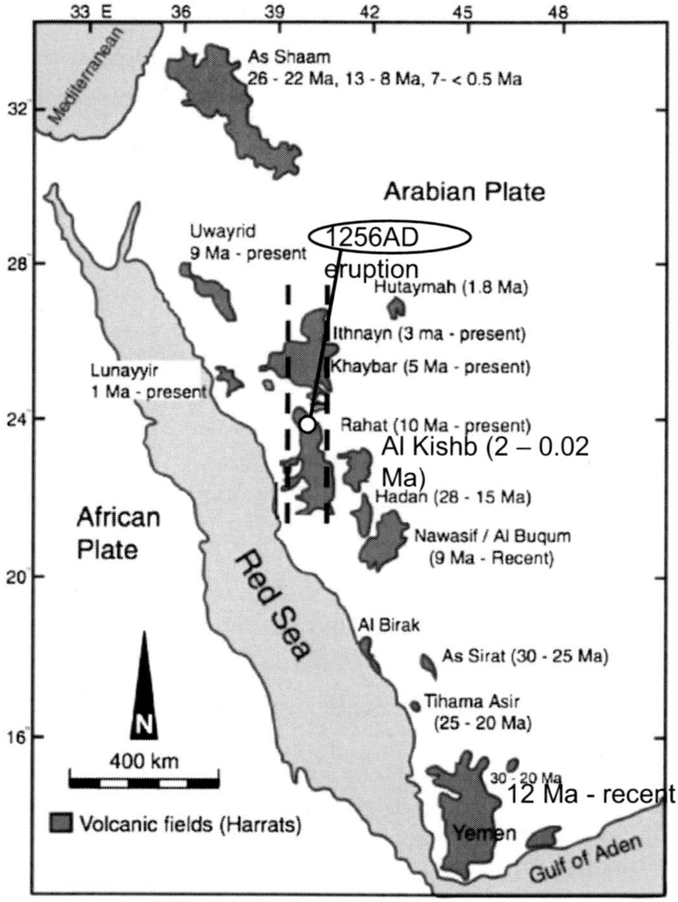

Figure 4. Harrats (volcanic fields) of the Arabian Peninsula and their eruption duration range (modified after Moufti et al., 2013). The white circle is the proposed "1256AD Al Madinah Historic Eruption Site". The dashed lines define the Makkah-Madinah-Nafud volcanic alignment (Camp et al., 1991).

3.2. Younger Harrats and Miocene-Recent Yemen Volcanic Series

Subsequently, volcanism ceased in most parts of the Red Sea basin. Exceptions are recorded in Afar, where a complex igneous history continued to the present day (Barberi et al., 1972; Varet 1978; Zanettin et al., 1978; Berhe 1986; Vellutini 1990), and at Harrat Ishara (north of Madinah), which experienced basaltic volcanism from ~17 to 14.5 Ma (Szymanski 2013). In the Middle Miocene, volcanism returned to far-reaching areas of the Arabian Peninsula: Harrat ash Shaam, Jordan (Ilani et al., 2001), the Younger Harrats of Saudi Arabia (Coleman 1993 and references therein), and the Miocene-Recent Yemen volcanic series (YVS Mattash et al., 2013 and references therein). The volcanism continues along the newly formed continental edge to the present day, albeit weaker than before.

3.2.1. The 1256 AD Al Madinah Historic Eruption

The eruption near the holy city of Al Madinah in 1256 AD is of particular historical importance being the last historically erupted volcano in the Kingdom of Saudi Arabia. It has been proposed as a major Geosite (Figure 4) named "*The 1256 AD Al Madinah Historic Eruption Site*" (Moufti et al., 2013; 2016).

The eruption occurred on the northern part of Harrat Rahat and formed a ~2.25 km long chain of NW–SE-aligned scoria and lava spatter cones, producing alkali-olivine basalt, "a'a" and "pahoehoe" lava flows. The event started on the 5th of June with an earthquake swarm. The strongest earthquake hit 4 days later and was followed by lava fountaining from a rift only 19 kilometres southeast of Al Madinah. Eyewitnesses gave a vivid account of the events, albeit paraphrased and reported 300 years later (Camp et al., 1987):

> "For days the volcanic eruption was preceded by many great earthquakes which occurred in Al Madinah at the beginning of Jumad-Al-Thani (Monday 1st of June 1256, A/N). At first the movements were

slight and not all of the residents of Madinah town felt them. On Tuesday, the second day of Jumad Al-Thani, the earthquakes became stronger. On Wednesday, in the third part of the night (the hours before sunrise, the night being divided into three parts, A/N), the greatest earthquake occurred, which frightened the residents. The earth tremors continued throughout the rest of the night. On Friday, a major event occurred, when the ground and ceiling of houses were shaken. Eighteen earthquakes were recorded during this day. [...] After the main earthquake was felt in Medinah at midday, fire appeared associated with black smoke clouds, which accumulated in the atmosphere. The greatest fire covered the horizon to the east of Madinah. [...] The lava flow carried along its way gravel, stones and trees. It was like a red-blue boiling river, with thundering noises. The lava flow moved toward the north of Wadi Eheline. The light of the fire was seen in Makkah, Busrah, and Taima. The historians wrote that the fire continued for three months [...] The lava flow descended on the rocky ground and was as high as a long spear (a couple of meters thick, A/N) above the ground level [...] When the lava flow came to a complete stop [...] it created a dam that formed a great lake in the rainy seasons (Al Habas Lake)."

Figure 5. Imaginative reconstruction of the 1256AD Al Madinah Historic Eruption.

4. Volcanism in the Red Sea trough

After continental rifting, oceanic spreading in the Red Sea occurred at about 5 Ma, with volcanic (mainly basaltic) activity (Cochran 1983; Bonatti 1985; Altherr et al., 1988, 1990; Bayer et al., 1988; Bohannon et al., 1989; Volker et al., 1993; Prodehl et al., 1997; Volker et al., 1997; Zeyen et al., 1997). Morphologically, two sublinear troughs form the Red Sea depression. The broad main trough extends from the Afar Triangle to the Gulf of Suez. Between 15°N and 19°N, the main trough is bisected by a narrow axial trough. North of 20°N the axial trough becomes discontinuous and there are some isolated deeps. Seismicity in the Red Sea axial zone is intense between 17°N and 19°N; south of 15°N it is displaced in northern Afar, while north of 20°N seismicity is nearly absent.

Three regions of magmatism can be characterized in the Red Sea trough (Figure 2).

4.1. Southern Red Sea Region

The axial zone of the southern Red Sea from about 15°N to 19°N displays a ~300 km long, nearly continuous, rift valley (Figure 2), whose floor reaches ~2,200 m below sea level. The spreading appears to be asymmetric at around 17°N, with rates about 70% higher on the western, African side. Basalts cover the seafloor of the valley and both seafloor magnetic anomalies and normal faulting earthquakes, typically associated with spreading ridges, are present. South of 16°N the axial trough becomes shallower, and three groups of islands emerge: Jabal at-Tair, Zubair group, and Hanish–Zukur group, with historical episodes of volcanic activity. The most recent episodes are represented by the eruption of 2007 on Jabal at-Tair (Mattash, 2008; Mattash et al., 2014, Xu and Jónsson, 2014), and by those of 2011 and 2013 in the Zubair group, where new islands emerged from the sea (Jónsson and Xu, 2015).

4.1.1. December 2011 - The Red Sea Has a New Inhabitant: A Smoking Island

"New Island Rises in the Red Sea" (by B. Israel, from the magazine "Live Science," 12/28/2011, https://www.livescience.com>Planet Earth): The island was created by a wild eruption that occurred in the Red Sea earlier this month. It is made of loose volcanic debris from the eruption, so it may not stick around long. According to news reports, fishermen witnessed lava fountains reaching up to 90 feet (30 meters) tall on Dec. 19, which is probably the day the eruption began. Ash plumes were seen emanating from the spot Dec. 20 and Dec. 22 by the Moderate Resolution Imaging Spectroradiometer (MODIS) on NASA's Terra and Aqua satellites. The Ozone Monitoring Instrument on NASA's Aura satellite detected elevated levels of sulphur dioxide, further indicating an eruption. By Dec. 23, what looked like a new island had appeared in the Red Sea off the west coast of Yemen.

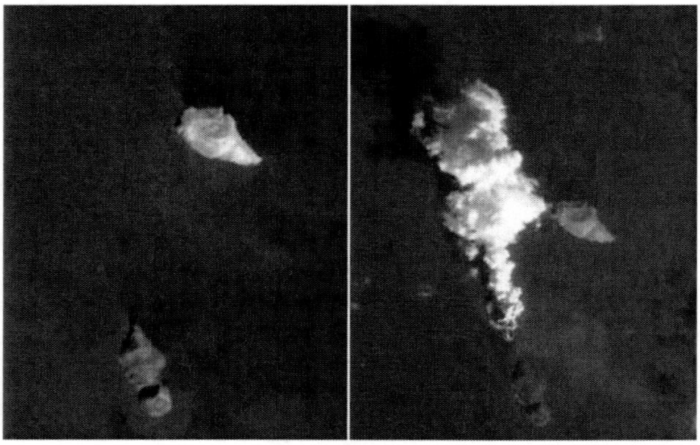

Figure 6. Before and after the volcano eruption - A photograph showing the Zubair Group of islands in October 2007, and right a photograph showing the islands on 23nd of December, 2011. Photo: AFP.

The volcanic activity occurred along the Zubair Group, a collection of small islands that run in a roughly northwest-southeast line. The islands rise from a shield volcano (a kind of volcano built from fluid lava flows) and poke above the sea surface.

4.2. Central Red Sea Region

Moving north from about 20°N, the axial zone of the Red Sea is characterized by a number of discontinuous deeps with a complex morphology and depths reaching over 2,000 m. They became the focus of much interest in the 1960s because it was found that they contain dense, hot brines as well as hydrothermal metalliferous deposits (Degens and Ross 1969). Some of the brines have temperatures >50°C. Deposits from the brines include abundant Fe-sulphides, often enriched in Zn and Cu, Fe-hydroxides (goethite), Fe-montmorillonite, Mn-oxides and hydroxides, manganosiderite and anhydrite (Bischoff 1969). Moving further north, a number of discrete axial troughs with high amplitude magnetic anomalies are separated by inter-trough zones that lack an axial rift valley and magnetic anomalies and have a thick continuous sediment cover across the axis (Figure 7). The width and length of the trough segments decrease moving north.

4.3. Zabargad and the Northern Red Sea

Next to Zabargad Island, the Red Sea axis is laterally displaced by about 100 km along a fracture zone (ZFZ, Figure 2) that can be considered the precursor to a major oceanic transform. The uplift of the island may be due in part to the transpression occurring along the SSW side of the fracture zone. Zabargad exposes an uplifted block of sub-Red Sea lithosphere, including upper mantle peridotites, gabbros, and granitoid gneisses criss-crossed by basaltic dykes, in addition to sandstone-limestone and evaporitic units (Bonatti et al., 1981, 1986; Bosworth et al., 1996). North of Zabargad, a few scattered deeps with basaltic and hydrothermal activity have been identified, whereas in two small islets ("The Brothers") gabbro outcrops were recorded.

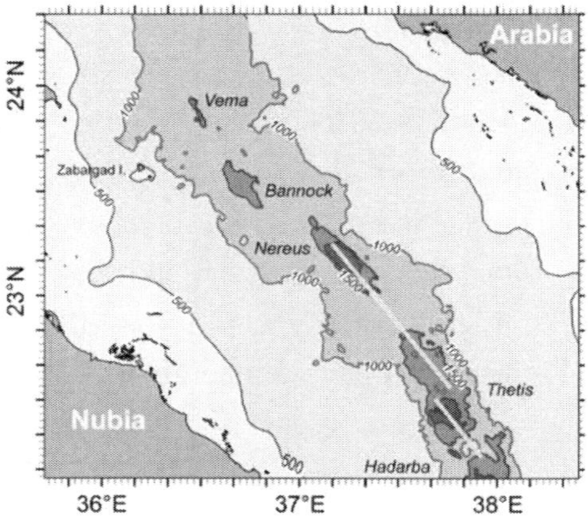

Figure 7. The central region of the Red Sea, with discrete axial rift valley segments separated by inter-trough zones (modified after Bonatti et al., 2015).

5. Chemical and Isotope Characteristics of the Red Sea Volcanism

In 1972, the R/V Glomar Challenger conducted Leg 23B in the Red Sea and drilled six wells (sites 225–230) in and nearby the axial trough (Whitmarsh et al., 1974). These wells confirmed that basalts and fluids with mantle-derived lead isotopes were emplaced into the sediment column along the Red Sea axis and supported the hypothesis that the Red Sea was an evolving oceanic rift. The subsequent ongoing exploration drilling programs produced a large amount of basaltic samples dredged from the sea floor, allowing several geochemical studies to be conducted all along the Red Sea (Altherr et al., 1988, 1990; Coleman and McGuire 1988; Eissen et al., 1989; Barrat et al., 1990; Rogers 1993; Volker et al., 1997; Haase et al., 2000; Ligi et al., 2012; Mattash et al., 2013, 2014, Pinarelli et al., 2015). A synthesis of the available chemical and isotope data is reported in Table 1.

Table 1. Chemical and Sr–Nd-Pb isotope compositions of selected samples of Red Sea basalts

Sample	SM126	Drl-1	Drl-3	Dr2-1	Dr2-2	Dr3-1	KL9-1	KL9-2	RS107	RM2	RM5	RM6	RM9	RM10	MR-83-3	MR-83-4	V84
Lat. N	26.14	26.12	26.12	26.13	26.13	26.14	26.15	26.15	25.33	25.33	25.33	25.33	25.33	25.33	23.13	23.13	21.25
Locality	Shaban Deep	Shaban Deep	Shaban Deep	Shaban Deep	Shaban Deep	Shaban Deep	Shaban Deep	Shaban Deep	Al Waijh Deep	Al Waijh Deep	Al Waijh Deep	Al Waijh Deep	Al Waijh Deep	Al Waijh Deep	Nereus Deep	Nereus Deep	Atlantis II Deep
SiO_2wt%	50.10	50.3	50.68	51.72	51.26	50.31	50.66	50.13	50.60	50.10	50.90	51.50	51.60	51.20	48.88	49.02	48.50
TIO_2	1.35	1.17	1.15	2.1	2.15	1.34	1.66	1.67	1.46	1.3	1.66	1.46	1.48	1.47	1.48	0.59	0.76
AL_2O_3	17.4	15.41	15.57	15.35	15.21	15.33	15.2	15.54	14.5	17.4	15.2	14.9	15	14.7	14.51	15.72	16
$Fe_2O_{3\ (tot)}$	8.50	9.26	9.12	10.71	10.46	9.06	9.37	9.61	11.38	9.72	11.77	11.50	11.20	11.50	13.00	9.64	9.75
MnO	0.13	0.13	0.15	0.18	0.2	0.15	0.18	0.15	0.18	0.14	0.17	0.18	0.18	0.19	0.19	0.23	0.16
MgO	7.31	7.83	7.82	6.05	5.97	7.53	7.18	7.21	7.03	6.04	5	6.92	6.6	6.92	8.36	9.97	10.6
CaO	11.6	12.17	12.21	9.72	9.68	12	11.79	12.21	11.2	13.1	11.8	11.6	11.7	11.5	11.54	13.81	12.4
Na_2O	3.04	3.1	3.21	3.81	3.64	3.16	3.42	3.42	2.77	2.93	3.22	3.05	2.92	2.94	2.34	1.71	2
K_2O	0.38	0.24	0.23	0.37	0.39	0.31	0.21	0.22	0.34	0.16		0.37	0.38	0.37	0.13		0.03
P_2O_5	0.16								0.15	0.09	0.14	0.14	0.13	0.13	0.02		0.06
Mg#	63.0	66.1	66.4	56.5	56.8	65.7	63.8	63.3	55.0	55.2	45.7	54.4	53.8	54.4	59.1	69.9	68.3
Cr ppm	305								54	139	52	56	48	51			467
Co	36	46	45	39	39	43	43	44	42	42	38	46	44	45			48
Ni	94	96	97	66	64	98	78	76	43	56	30	42	44	45			229
Sc	33	36	34	35	34	36	52	42	46	42	45	45	46	45			38
Rb	13.0	5.8	5.4	8.5	8.3	6.9	4.8	4.7	11.0	3.0	12.0	11.0	10.0	11.0			0.5
Cs		0.11	0.10	0.15	0.14	0.10	0.09	0.11									
Sr	197	145	137	185	181	158	191	196	128	142	134	128	135	132			63
Ba	63	55	55	64	65	62	39	40	67	52	86	87	94	70			
Zr	96	56	53	189	188	76	108	109	94	82	100	90	93	92			34
Hf	2.3	1.8	1.7	4.9	4.8	2.1	2.7	2.7	2.5	1.8	2.5	2.2	2.0	2.2			1.1
Nb	10	6	5	8	7	7	5	4	10	5	9	9	9	9			1

Table 1. (Continued)

Sample	SM126	Drl-1	Drl-3	Dr2-1	Dr2-2	Dr3-1	KL9-1	KL9-2	RS107	RM2	RM5	RM6	RM9	RM10	MR-83-3	MR-83-4	V84
Lat. N	26.14	26.12	26.12	26.13	26.13	26.14	26.15	26.15	25.33	25.33	25.33	25.33	25.33	25.33	23.13	23.13	21.25
Locality	Shaban Deep	Shaban Deep	Shaban Deep	Shaban Deep	Shaban Deep	Shaban Deep	Shaban Deep	Shaban Deep	Al Waijh Deep	Al Waijh Deep	Al Waijh Deep	Al Waijh Deep	Al Waijh Deep	Al Waijh Deep	Nereus Deep	Nereus Deep	Atlantis II Deep
Ta ppm	0.49	0.36	0.36	0.50	0.52	0.48	0.31	0.31	0.52	0.27	0.44	0.44	0.47	0.45			0.04
Th	1.02	0.48	0.49	0.63	0.63	0.59	0.37	0.39	0.60								0.05
U	0.30	0.12	0.17	0.18	0.19	0.16	0.13	0.15	0.21		0.50	0.36	0.57				0.02
Y	29	23	22	47	46	24	30	30	33	28	39	34	35	34			
La	6.4	4.3	4.4	8.4	8.4	5.6	5.1	5.2	5.1	2.9	6.1	5.9	5.5	5.5			0.5
Ce	13.8	10.1	10.3	24.0	23.8	13.1	14.3	14.6	13.4	8.8	14.4	13.6	13.8	12.9			2.6
Pr		1.51	1.55	3.89	3.83	1.91	2.33	2.35									
Nd	10.7	7.8	7.8	20.0	19.4	9.5	11.7	11.9	10.5	7.4	11.2	10.0	9.7	9.2			2.7
Sm	3.1	2.6	2.6	5.9	6.0	2.8	3.7	3.6	3.6	3.1	4.2	3.7	3.6	3.6			1.6
Eu	1.05	0.96	0.99	1.92	1.87	1.07	1.33	1.37	1.47	1.05	1.30	1.18	1.15	1.13			0.74
Gd		3.2	3.3	6.8	6.9	3.6	4.5	4.5	5.6	4.1	5.7	5.0	5.1	4.8			2.9
Tb	0.61	0.57	0.58	1.19	1.21	0.62	0.76	0.77	0.90	0.77	1.09	0.89	0.88	0.80			0.51
Dy		3.8	3.9	7.9	7.8	4.0	5.2	5.1		4.8		5.3	4.9				
Ho	0.97	0.79	0.82	1.62	1.63	0.84	1.04	1.06	1.30	1.20	1.70	1.20	1.30	1.30			0.80
Er		2.3	2.4	4.6	4.7	2.5	3.0	3.0									
Tm	0.38	0.33	0.34	0.68	0.66	0.35	0.42	0.42	0.53	0.42	0.56	0.49	0.44	0.46			0.36
Yb	2.4	2.1	2.2	4.4	4.3	2.3	2.7	2.8	3.5	2.6	3.5	3.0	3.0	3.0			2.3
Lu	0.37	0.31	0.32	0.66	0.64	0.34	0.40	0.41	0.56	0.38	0.51	0.48	0.47	0.45			0.35
$^{87}Sr/^{86}Sr$	0.703080	0.703147	0.703148	0.703077	0.703046	0.703134	0.703083	0.703089	0.703121	0.703330	0.703240	0.703210	0.703110	0.703120	0.702980	0.702890	0.702692
$^{143}Nd/^{144}Nd$		0.513076	0.513072	0.513098	0.513093	0.513088	0.513110	0.513114	0.513070						0.513150	0.513100	0.513154
$^{206}Pb/^{204}Pb$	18.269								18.859								18.114
$^{207}Pb/^{204}Pb$	15.595								15.561								15.440
$^{208}Pb/^{204}Pb$	37.979								38.770								37.875
Reference	1;2	3	3	3	3	3	3	3	1	1;2	1;2	1;2	1;2	1;2	5	5	1;4;6

Sample	23/B	V1G	V19	V73	17/1385	17/1388	V62	RS52	RS37	RS43U	59-3	53-2	68-2	71-2	17/1389	V43	YE2
Lat. N	21.25	21.25	21.08	20.07	20.01	20.01	19.64	19.07	18.15	18.15	17.58	17.57	17.57	17.57	17.00	16.00	15.33
Locality	Atlantis II Deep	Atlantis II Deep	Shagara Deep	Port Sudan Deep	Volcano Deep	Volcano Deep	Suakin Deep	Axial Trough	Axial Trough	Axial Trough	Axial Trough	Axial Trough	Axial Trough	Axial Trough	Ramad seamount	Axial Trough	Jabal at Tair Island
SiO_2wt%	49.60	48.70	49.70	49.80	49.80	49.90	49.40	50.00	50.50	50.90	50.52	50.62	50.68	50.88	47.40		50.20
TIO_2	1.28	1.01	1.29	1.15	1.49	1.52	1.48	1.66	1.47	1.51	0.88	1.15	1.5	1.1	1.59		1.93
AL_2O_3	14.9	15.9	14.3	14.2	14.4	14.4	14.1	13.3	13.6	13.7	14.3	14.1	14.4	14.2	16.2		14.18
$Fe_2O_{3(tot)}$	12.00	10.57	12.18	12.42	12.58	12.65	13.66	15.79	14.24	14.40	11.32	12.72	12.92	12.77	9.50		13.00
MnO	0.2	0.16	0.19	0.19	0.19	0.19	0.2	0.23	0.21	0.21	0.18	0.19	0.19	0.2	0.14		0.21
MgO	7.61	9.45	8.04	9.4	7.34	7.37	7.4	6.94	6.82	6.94	8.3	7.42	7.1	7.29	8.7		6.26
CaO	12.2	12.5	12	11.5	11.8	11.7	11.3	10.4	10.7	10.7	12.8	12.15	10.8	11.72	12.7		10.87
Na_2O	2.6	2.28	2.42	2.05	2.37	2.36	2.27	2.34	2.2	2.35	1.92	1.91	2.28	1.97	2.6		2.94
K_2O	0.14	0.06	0.09	0.15	0.19	0.18	0.21	0.16	0.18	0.18	0.07	0.16	0.18	0.2	0.69		0.37
P_2O_5	0.1	0.07	0.12	0.12	0.17	0.17	0.17	0.16	0.16	0.15	0.04	0.1	0.08	0.09	0.29		0.22
Mg#	55.7	63.9	56.7	60.0	53.6	53.6	51.8	46.5	48.7	48.8	59.2	53.6	52.1	53.1	64.5		48.8
Cr ppm	226	431	242	388	289	282	266	224	134	134					292		
Co	46	51	46	54	47	46	46	62	54	56					41	45	
Ni	80	187	103	172	100	106	105	95	66	69					128	34	17
Sc	42	39	47	43	44	44	45	47	49	45					37	39	
Rb	1.5	2.0	1.6	3.1	5.0	5.0	4.8	5.0	6.0	6.0					16.0	8.4	6.8
Cs																	
Sr	90	104	73	63	98	98	84	81	91	91					332		201
Ba	16	11			27	30		33	40	42					190	111	72
Zr	16	72			106	107	100	103	95	96					142	169	114
Hf	2.2	1.4	2.4	1.7	2.3	2.3		2.9	2.1	1.0					2.7	3.4	
Nb ppm		2			6	6	7	7	9	9					22		14
Ta	0.10	0.05	0.15	0.30	0.34	0.34	0.46	0.39	0.47	0.43					1.50	1.22	0.90
Th				0.50	0.30	0.31	0.67		0.48	0.35					1.74	1.35	1.10
U			0.29	0.09			0.26								0.63	0.55	
Y		28			38	39		46	36	37					24		
La	1.8	1.0	2.6	3.5	4.4	4.6	5.3	4.4	5.3	5.2					22.6	12.2	9.8

Table 1. (Continued)

Sample	23/B	V1G	V19	V73	17/1385	17/1388	V62	RS52	RS37	RS43U	59-3	53-2	68-2	71-2	17/1389	V43	YE2
Lat. N	21.25	21.25	21.08	20.07	20.01	20.01	19.64	19.07	18.15	18.15	17.58	17.57	17.57	17.57	17.00	16.00	15.33
Locality	Atlantis II Deep	Atlantis II Deep	Shagara Deep	Port Sudan Deep	Volcano Deep	Volcano Deep	Suakin Deep	Axial Trough	Axial Trough	Axial Trough	Axial Trough	Axial Trough	Axial Trough	Axial Trough	Ramad seamount	Axial Trough	Jabal at Tair Island
Ce	6.5	4.6	7.5	9.2	11.0	10.6	13.5	11.7	11.5	12.8					38.5	29.2	24.7
Pr																	3.31
Nd	7.1	5.1	6.4	7.6	7.9	8.1	9.7	9.8	9.6	9.5					18.3	17.8	17.3
Sm	3.0	2.3	2.7	2.4	3.3	3.4	3.1	3.8	3.5	3.2					4.6	4.6	4.3
Eu	1.00	0.85	1.12	0.95	1.00	1.02	1.26	1.46	1.36	1.20					1.25	1.60	1.51
Gd	4.1	2.9		3.6	4.2	4.4	4.8	6.1	1.2	4.2						5.2	5.0
Tb	0.70	0.58	0.77	0.68	0.74	0.76	0.92	1.16	0.96	0.83					0.60		0.91
Dy																5.6	5.4
Ho	1.00	0.80	1.20	1.07	1.24		1.36	1.65	1.42	1.17					0.79		1.11
Er																3.5	3.3
Tm	0.46	0.35	0.57		0.50	0.47	0.67	0.90	0.60	0.57							0.50
Yb	3.0	2.4	3.6	3.4	3.7	3.6	4.1	5.6	3.8	4.0					2.0	3.0	3.1
Lu	0.48	0.43	0.54	0.50	0.55		0.61	0.82	0.60	0.55						0.45	0.47
$^{87}Sr/^{86}Sr$	0.702730	0.702691	0.702520	0.702973	0.703020	0.702980	0.702944	0.702950	0.702913	0.702891	0.702861	0.702906	0.702914	0.702906	0.703178		
$^{143}Nd/^{144}Nd$	0.513194	0.513142		0.513079			0.513070	0.513090	0.513063	0.513062	0.513056	0.512990	0.513020	0.513014	0.512968	0.513002	
$^{206}Pb/^{204}Pb$	18.672	18.267		18.788			18.829			18.976	18.787	19.116	19.141	19.163	19.597		
$^{207}Pb/^{204}Pb$	15.543	15.484		15.546			15.551			15.562	15.555	15.590	15.577	15.588	15.646		
$^{208}Pb/^{204}Pb$	38.307	38.058		38.519			38.557			38.726	38.575	38.857	38.876	38.895	39.507		
Reference	1;2	1;4	1;2	1;3	1;2	1;2	1;4	1;4	1;4	1;4	7	7	7	7	1;4	6	9

Sample	YE3	YE4	YE5	YE6	Z4	Z10	Z2	Z13	JZ9A	JZ11	Z48	H18	H4	H24	H30	H42
Latitude N	15.33	15.33	15.33	15.33	15.10	15.10	15.10	15.10	15.10	15.10	15.10	13.45	13.45	13.45	13.45	13.45
Locality	Jabal at Tair Island	Jabal at Tair Island	Jabal at Tair Island	Jabal at Tair Island	Zubair Islands	Zubair Islands	Zubair Islands	Zubair Islands	Zubair Islands	Zubair Islands	Zubair Islands	Hunaish Islands	Hunaish Islands	Hunaish Islands	Hunaish Islands	Hunaish Islands
SiO_2 wt%	50.30	49.30	48.40	49.40	47.90	48.00	47.90	45.85	47.84	47.50	48.00	47.80	45.75	46.29	46.89	44.00
TIO_2	2	2.12	1.94	1.29	2.81	2.03	2.06	1.94	1.82	1.81	2.01	2.14	2.87	2.08	2.11	2.82
AL2O3	14.11	13.64	14.86	17.44	16.1	15.3	15.34	14.5	16.85	16.39	15.3	15.5	19.28	14.94	15.1	18.36
$Fe_2O_{3\,(tot)}$	13.40	13.40	12.50	9.40	10.94	10.10	10.08	9.37	9.70	9.76	10.30	10.92	11.08	10.60	10.75	10.91
MnO	0.22	0.21	0.2	0.15	0.16	0.15	0.15	0.14	0.15	0.15	0.15	0.17	0.16	0.14	0.14	0.16
MgO	6.12	5.56	5.03	6.19	6.32	9.9	10.28	9.5	7.52	7.79	10	9.38	4.64	9.72	10.19	4.54
CaO	10.6	10.54	11.23	12.69	11.2	10.1	9.94	10.55	11.8	11.99	10.1	10.1	11.69	9.29	9.43	11.39
Na_2O	2.99	3.05	3.19	2.44	3.33	3.19	3.17	3.68	2.66	2.7	3.15	3.04	3.06	2.69	2.67	3.95
K_2O	0.36	0.4	0.41	0.29	0.81	0.76	0.71	0.47	0.48	0.48	0.75	0.62	0.89	0.56	0.56	0.62
P_2O_5	0.25	0.26	0.25	0.17	0.51	0.42	0.4	0.58	0.27	0.27	0.4	0.31	0.43	0.3	0.3	0.43
Mg#	47.5	45.1	44.3	56.6	53.2	65.6	71.6	71.5	65.7	64.4	65.8	63.0	50.9	69.4	70.1	50.7
Cr ppm					153	502	523	485	285	292	504	482	54	443	436	55
Co					37	44	45	42	39	40	45	52	34	48	48	35
Ni	13	15	15	24	70	259	244	223	117	134	263	511	20	227	227	20
Sc					32	28	28	29	34	40	29	29	25	30	29	25
Rb	7.2	5.7	6.6	5.8	17.0	16.0	18.0	12.0	8.0	10.0	13.0	12.0	20.0	13.0	11.0	18.0
Cs																
Sr	210	229	245	229	446	393	424	406	376	361	392	364	625	337	336	610
Ba	76	82	84	62	217	210					187	150				
Zr	121	129	121	79	217	188	181	165	138	147	188	152	185	158	155	171
Hf					4.8	4.1	3.9	3.7	3.1	3.4	4.1	3.8	4.2	3.6	3.6	4.2
Nb	14	15	14	10	37	32	32	28	21	22	31	22	36	23	21	34
Ta	1.00	0.90	0.90	0.80	2.08	1.65	2.07	1.97	1.30	1.45	1.68	1.39	2.48	1.40	1.36	2.47

Table 1. (Continued)

Sample	YE3	YE4	YE5	YE6	Z4	Z10	Z2	Z13	JZ9A	JZ11	Z48	H18	H4	H24	H30	H42
Latitude N	15.33	15.33	15.33	15.33	15.10	15.10	15.10	15.10	15.10	15.10	15.10	13.45	13.45	13.45	13.45	13.45
Locality	Jabal at Tair Island	Jabal at Tair Island	Jabal at Tair Island	Jabal at Tair Island	Zubair Islands	Zubair Islands	Zubair Islands	Zubair Islands	Zubair Islands	Zubair Islands	Zubair Islands	Hunaish Islands	Hunaish Islands	Hunaish Islands	Hunaish Islands	Hunaish Islands
Th ppm	1.20	1.20	1.40	1.00	2.35	2.05	0.32	2.76	1.79	1.94	2.03	1.06	2.94	1.98	1.78	3.02
U					0.75	0.71					0.71	0.41				
Y					31	27	27	24	22	27	27	30	26	26	26	24
La	11.2	11.3	11.1	8.7	30.4	24.2	27.7	24.1	15.3	17.2	25.5	18.3	35.4	16.9	16.3	25.7
Ce	27.6	28.5	27.8	21.1	58.0	45.0	49.8	49.4	31.0	39.4	42.7	38.0	53.8	36.5	34.8	53.2
Pr	3.69	3.72	3.66	2.67												
Nd	16.9	18.9	16.7	11.6	37.3	27.2		26.5	17.7	21.5	26.9	21.9	28.7	22.5	20.7	30.5
Sm	4.4	4.8	4.3	3.0	7.1	5.5	5.5	5.2	4.0	4.9	5.5	5.3	6.4	5.2	4.9	6.4
Eu	1.65	1.65	1.56	1.17	2.01	1.58	1.86	1.82	1.50	1.71	1.60	1.59	2.21	1.86	1.81	2.20
Gd	5.5	5.6	5.2	3.4	5.7	4.4					4.1	5.1				
Tb	1.00	1.00	0.95	0.60	0.89	0.69	0.84	0.92	0.71	0.91	0.72	0.74	1.09	0.93	0.84	1.17
Dy	6.0	5.9	5.5	3.6												
Ho	1.24	1.17	1.09	0.73	0.97						0.77	0.76				
Er	3.4	3.3	3.2	2.0												
Tm	0.51	0.51	0.49	0.32												
Yb	3.4	3.2	3.0	2.1	2.6	2.2	2.2	2.0	1.8	2.3	2.2	1.9	2.1	2.1	2.0	2.1
Lu	0.49	0.48	0.46	0.30	0.38	0.36	0.32	0.30	0.29	0.35	0.29	0.26	0.31	0.33	0.31	0.33
$^{87}Sr/^{86}Sr$					0.703311	0.703320	0.703430	0.703800	0.703690	0.703590	0.703285	0.703204	0.703310	0.703230	0.703240	0.703960
$^{143}Nd/^{144}Nd$					0.512953	0.512958	0.513075	0.513013	0.513026	0.513092	0.512970	0.512997	0.513060	0.513035	0.513055	0.513077
$^{206}Pb/^{204}Pb$					19.130	18.857	19.041	19.022		19.077	19.113	19.858		18.675	18.788	19.002
$^{207}Pb/^{204}Pb$					15.565	15.652	15.547	15.567		15.694	15.573	15.540		15.505	15.512	15.539
$^{208}Pb/^{204}Pb$					39.034	38.721	38.855	38.841		39.124	39.010	38.737		38.460	38.582	38.762
Reference	9	9	9	9	10	10	11	11	11	11	10;12	10;12	11	11	11	11

(1) Altherr et al., 1988; (2) Altherr et al., 1990; (3) Haase 2000; (4) Volker et al., 1993; (5) Antonini et al., 1998; (6) Barrat 1990; (7) Eissen et al., 1989; (8) Volker et al., 1997; (9) Mattash 2014; (10) Altherr 1988; (11) Rogers 1993; (12) Volker 1997.

5.1. Regional Variations along the Red Sea

The lavas erupted in the Red Sea trough are olivine tholeiites with low K_2O (mostly <1%, Figure 8a) and low TiO_2 (mostly ≤2.0%, Figure 8b). The floor basalts, however, show significant chemical fluctuations.

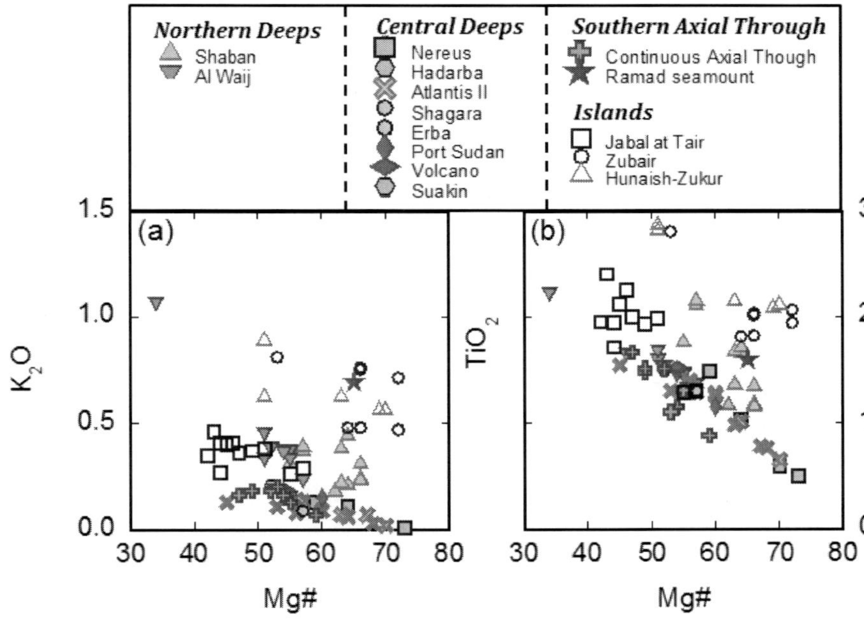

Figure 8. (a) K_2O and (b) TiO_2 vs. Mg# (=100xMgO/(MgO+FeO*)) in basalts from the Red Sea (data from Table 1).

To identify the main factors which control chemical variability in these magmas, the highly "hygromagmatophile" elements (such as Nb, Ta, Rb, Ba, Tb, Th, and U) can help a great deal since they are weakly or not fractionated during crystal fractionation processes (Joron and Treuil, 1989). In the La vs. Ta diagram (Figure 9a), all the samples follow the same correlation, with restricted fluctuations of the La/Ta ratio between 14 and 10. The latter value is characteristic of the spreading-centre domains. In a diagram that plots two different hygromagmatophile element ratios one versus the other, samples originating from chemically different mantle sources would plot along distinct correlation lines. For instance in the

Th/Ta vs. Th/Tb diagram (Figure 9b), values of the (Th/Ta)/(Th/Tb) ratio around 1 are typical of oceanic floor basalts. In this graphic we can observe three, possibly four, main trends showing that the Red Sea basalts may represent melts derived from compositionally distinct mantle sources.

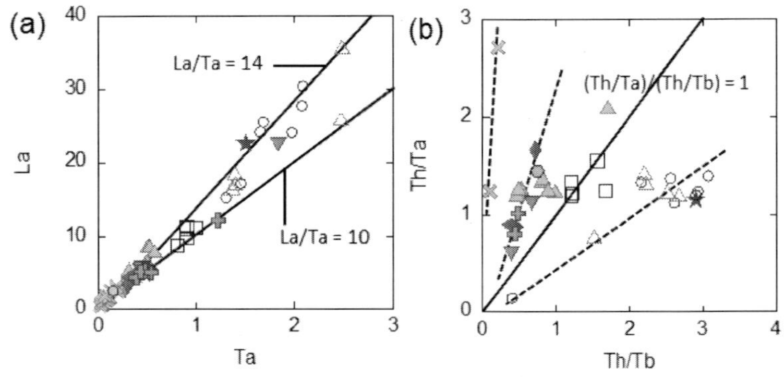

Figure 9. (a) La vs. Ta and (b) Th/Ta vs. Th/Tb in basalts from the Red Sea (data from Table 1). Symbols as in Figure 8.

It is worth noting that the chemical variations in floor basalts follow the geographical position along the Red Sea. Indeed, starting from the central deeps, the basalts display variable increases of K_2O, Rb, Sr, Ba, Nb, La, and Ta towards both north and south (Figure 10). The most primitive chemical compositions, with the lowest K_2O and incompatible trace element contents, are observed in the Deeps of the central region. Starting from here, the contents of incompatible elements increase in the basalts from both northern and southern outcrops. As regards the southern region, the contents of incompatible elements increase until Ramad seamount (17°N), but at the island of Jabal at-Tair they sharply decrease. Further south, the contents of incompatible elements resume to increase at Zubair and Hunaish-Zukur islands. In the latter basalts, the enrichment in K_2O causes some samples to assume an alkaline character (13a).

The variations described above are also detectable in the primitive mantle normalised incompatible element patterns (Figures 11 and 12). In this kind of diagrams (called "spidergrams"), the mantle components display increasing contents of large ion lithophile elements (LILE: Rb, Th, U, LREE, etc.) with respect to high field strength elements (HFSE: Ta, Zr, Hf), moving from the typical mantle source of oceanic basalts (Normal-MORB, N-MORB) up to HIMU (High U/Pb ratio; see § 5.3 for explanation). The higher the LILE/HFSE ratios, the more "enriched" is defined the mantle component.

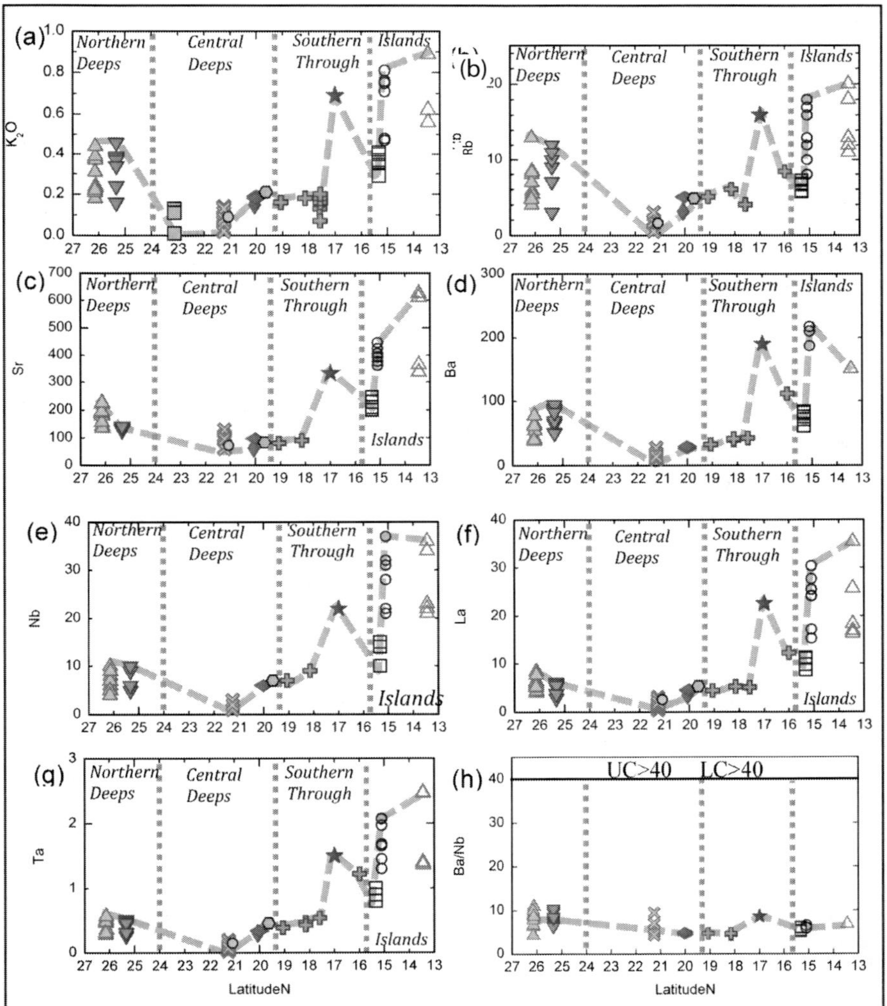

Figure 10. Chemical variations versus the Latitude N in the basalts from the Red Sea (data from Table 1). Symbols as in Figure 8. (h) UC = Upper Crust, LC = Lower Crust, BCC = Bulk Continental Crust (Rudnick and Gao, 2003).

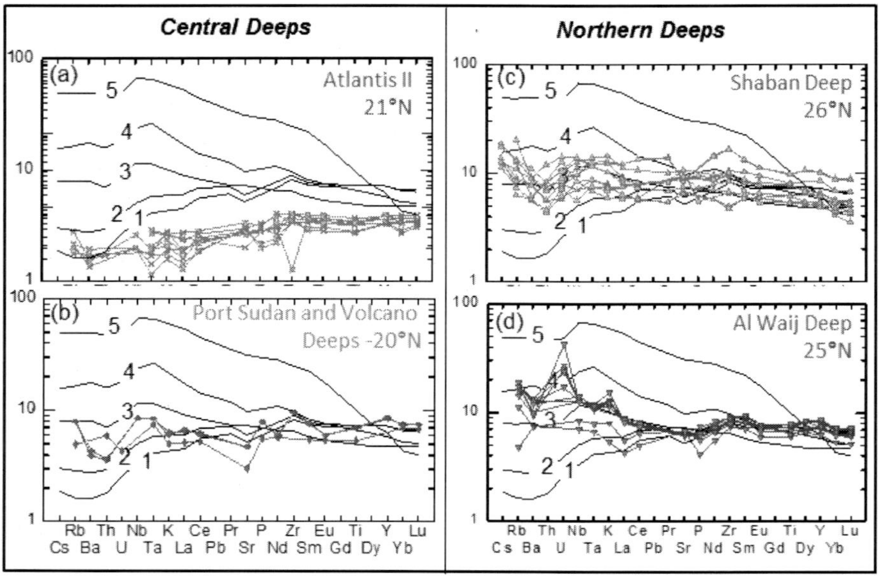

Figure 11. Primitive mantle normalised incompatible element patterns ("spidergrams") of the basalts from the Red Sea northern and central deeps (data from Table 1). The classical mantle end members are also represented: 1 = N-MORB; 2 = E-MORB; 3 = EMI; 4 = EMII; 5 = HIMU (data from Stracke et al., 2003; Willbold and Stracke, 2006; Jackson and Dasgupta, 2008; Gale et al., 2013). Normalising values after Sun and McDonough (1989).

The patterns of the central deeps are as primitive as those of either N-MORB (Atlantis II, Figure 11a), or E-MORB (Port Sudan and Volcano, Figure 11b). The northern deeps (Figure 11c and d), instead, show higher LILE/HFSE, with a shift of the patterns toward those typical of enriched mantle (EMI and EMII).

The basalts of the southern axial trough also show higher LILE contents with respect to the central deeps, with spidergrams similar to those of EMI (Figure 12a). Incompatible elements reach maximum contents at Ramad, which pattern approaches the even more enriched mantle component HIMU (Figure 12b). As already seen from Figure 10, a reversal of the increase of LILE towards south takes place on the island of Jabal at-Tair (Figure 12c). Further south, the Zubair and Hunaish Island groups (Figure 12d and e) return to exhibit enriched patterns, approaching the HIMU configuration.

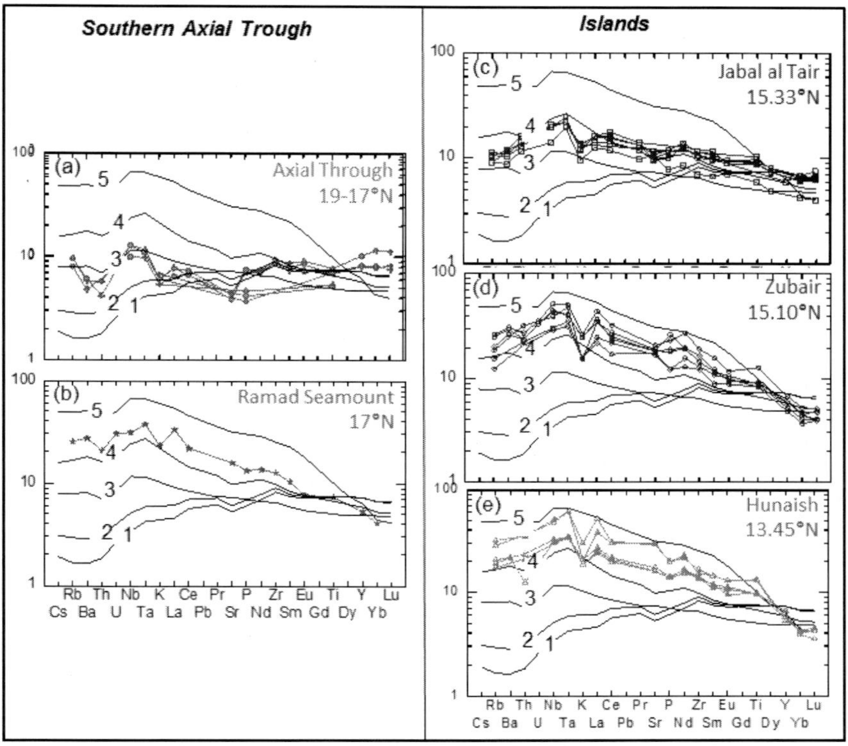

Figure 12. Primitive mantle normalised incompatible element patterns ("spidergrams") of the basalts from the Red Sea southern axial trough and islands (data from Table 1). The classical mantle end members are also represented: 1 = N-MORB; 2 = E-MORB; 3 = EMI; 4 = EMII; 5 = HIMU (data from Stracke et al., 2003; Willbold and Stracke, 2006; Jackson and Dasgupta, 2008; Gale et al., 2013). Normalising values after Sun and McDonough (1989).

5.2. The 2007 Eruption on Jabal at-Tair Island

The island of Jabal at-Tair is a key-outcrop in the context of the regional geochemical variations along the Red Sea. Indeed, it represents a reversal of the general increase of "enriched" geochemical characteristics towards south, with the reprise of more primitive features (Figures 10 and 12), which then continue to increase further south.

Jabal at-Tair is an oval-shaped island of about 9 km^2, composed of thin basaltic mainly "pahoehoe" and "aa" lava flows, as well as pyroclastic materials. On September 30, 2007, Jabal at-Tair renewed its volcanic activity, with Hawaiian and Strombolian eruptions, accompanied by abundant lava flows.

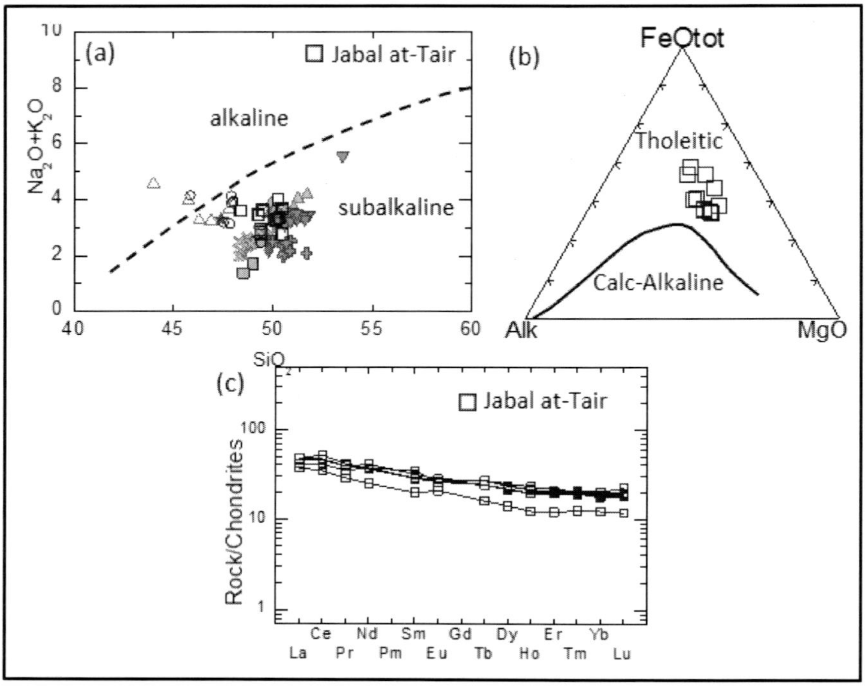

Figure 13. (a) Total alkalis vs. silica (TAS) classification diagram (after LeBas et al., 1986) plotting the basalts from the Red Sea (data from Table 1). Symbols as in Figure 8. (b) AFM plot of the Jabal at-Tair basalts. Tholeiitic/Calc-Alkaline divide after Irvine and Baragar (1971). (c) Chondrite normalised REE patterns of Jabal at-Tair basalts. Chondrite values after Sun and McDonough (1989).

The newly erupted lavas are geochemically similar to the older ones (Mattash et al., 2014). All of them have subalkaline character (Figure 13a) and show FeO enrichments typical of the tholeiitic trend (Figure 13b).

Their chondrite-normalized Rare Earth Elements (REE) patterns are considerably homogeneous (Figure 13c), with slight enrichments of light REE and slight depletions of heavy REE with respect to chondrite, resulting in moderate negative fractionations, and no Eu anomalies.

5.3. Origin of the Enriched Component

The primitive chemical composition of the central deeps are typical of a MORB-type mantle source, whereas the increase of K_2O and incompatible elements in the basalts from both southern and northern regions, could imply either crustal contamination or mixing with a mantle source having "enriched" signatures. Crustal contamination can be excluded on the basis of specific element/element ratios, which are extremely high in crustal rocks with respect to mantle-derived rocks. For instance Ba/Nb (Figure 10h) is higher than 40 in upper, lower, and bulk continental crustal rocks, whereas it is poorly variable and lower than 10 in the Red Sea basalts. Then, the variable enrichment in K_2O and incompatible elements of the Red Sea basalts in both northern and southern sectors can be ascribed to variable mixing with an "enriched" mantle source.

There are various sources identified as "enriched" mantle, which are inferred from differences in radiogenic isotope ratios that magmas inherit from their source rock (White, 1985; Zindler and Hart, 1986; Hart et al., 1992; Stracke et al., 2003, 2005). The sources, as defined by Sr, Nd and Pb isotopes, are as follows: (i) EMI (Enriched Mantle I): mantle contaminated with material derived from subducted pelagic sediments. (ii) EMII (Enriched Mantle II): mantle contaminated with material derived from the recycling of terrigenous sediments. (iii) HIMU (High U/Pb ratio): mantle contaminated by subducted oceanic crust that has not been homogenized with the rest of the mantle.

Magma genesis in the Afro-Arabian rift system has been ascribed to the contribution of a wealth of source domains, including a HIMU (Schilling et al., 1992; Deniel et al., 1994; Pik et al., 1999) or FOZO (Focus Zone: Marty et al., 1993, 1996; Orihashi et al., 2001, Mattash et al, 2014) component, a depleted (N-MORB) mantle reservoir, and both EMI–EMII components (Hart et al., 1989; Schilling et al., 1992; Chazot and Bertrand 1993; Orihashi et al., 2001; Furman et al., 2006, Mattash et al, 2014).

As regards the Red Sea basalts (Figure 14), each of the Sr, Nd and Pb isotope systematics provides peculiar evidence about the mantle sources. Sr isotopes (Figure 14a) are compatible with both MORB and HIMU components in most of the basalts, but those from the southern islands, which imply the contribution of an EMI source. Nd isotopes (Figure 14b) are definitely indicative of MORB-type mantle, but those of Ramad seamount and some of the Zubair Islands, which plot in the HIMU field.

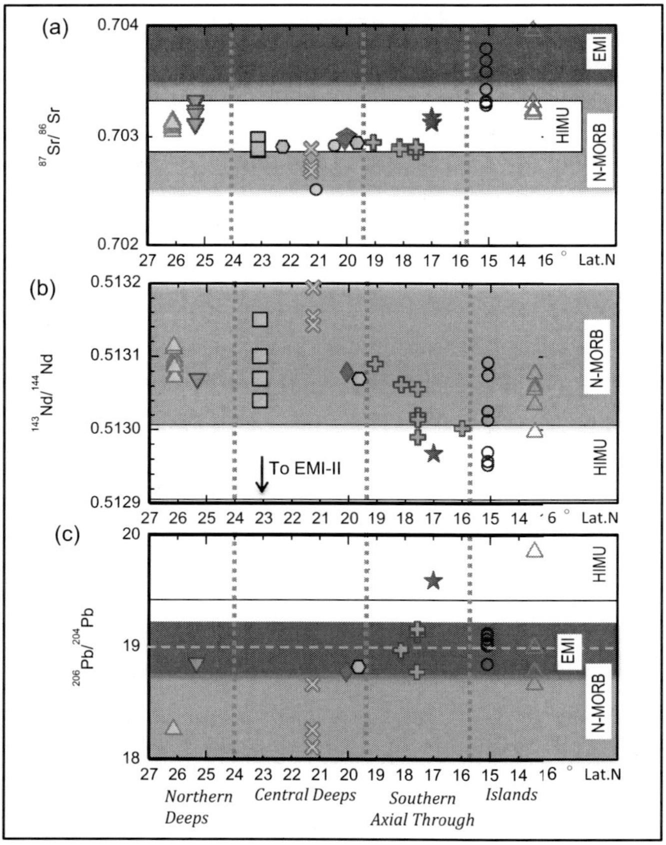

Figure 14. (a) Sr, (b) Nd and (c) Pb isotope variations versus Latitude N in the Red Sea basalts (data from Table 1). Symbols as in Figure 8. The plot shows the ranges of published data for mantle components: 1 = N-MORB; 2 = E-MORB; 3 = EMI; 4 = EMII; 5 = HIMU (Sun and McDonough 1989; Halliday et al., 1995; Willbold and Stracke 2006).

Finally, Pb isotopes (Figure 14c) show values mostly in the fields of either MORB or EMI, but those of Ramad seamount and one sample from Hunaish Islands, which denote the input of a HIMU source.

Combining the indications derived from the Sr, Nd and Pb isotope systematics, a three component system can be envisaged: MORB (central deeps, all the three systematics) plus HIMU (Nd and Pb isotopes of Ramad, Figure 14c) plus EMI (Sr isotopes of southern islands, Figure 14a). These mantle sources were likely mixed at variable degrees and extents to produce the relatively large variations in terms of isotope and chemical compositions observed in the Red Sea basalts.

Conclusion

The Red Sea is part of a continent scale rift system located along a divergent plate boundary that divide the African, Somalian, and Arabian plates.

In the late Oligocene-early Miocene through to the present time, large volumes of flood basalts (Harrats) emplaced along the western margin of the Arabian plate from the Gulf of Aden to the Mediterranean area. This magmatism started at ~31 Ma and continued to 16 Ma, with emplacements of large volumes of flood basalts in Afar, Arabia (Older Harrats) and Yemen (Trap basalts) (Figure 15a). The western and southern margins of the Arabian Plate were uplifted, creating the present Red Sea Escarpment and the mountains beyond. Continental rifting that caused the separation of the African plate from that of Arabia started at ~24 Ma (Figure 15b). Afterward, the volcanism ceased in most parts of the Red Sea basin until Middle Miocene, when Younger Harrats in Arabia and Miocene-Recent volcanic series in Yemen were emplaced. The volcanism has continued along the newly formed continental edge, and still continues today, albeit weaker than before. Oceanic spreading followed advanced continental rifting at ~5 Ma (Figure 15c), with volcanic activity in the Red Sea trough.

The Red Sea displays a nearly continuous rift valley in its southern part, from about 15°N to 19°N, which encompasses the place of the mid-ocean ridge initiation (Figure 15c). South of 16°N this axial trough becomes shallower, and three groups of islands emerged: Jabal at-Tair, and the Zubair and Hanish–Zukur groups, with historical episodes of volcanic activity.

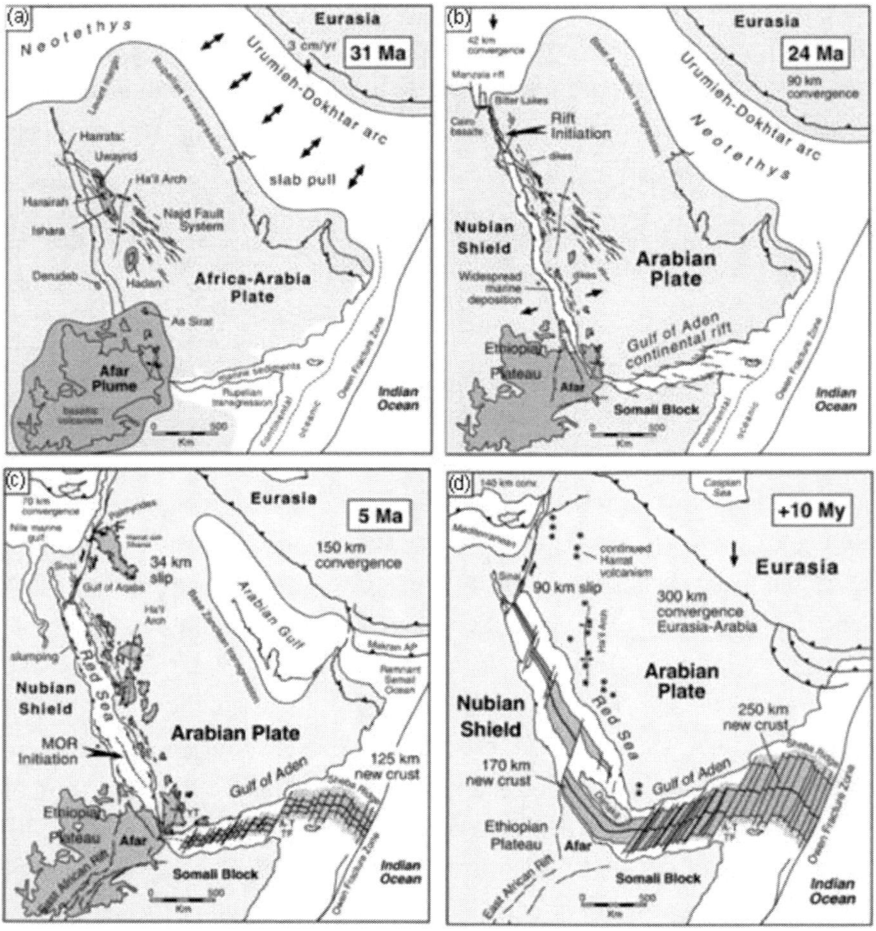

Figure 15. Palinspastic restoration of the Red Sea–Gulf of Aden rift system (modified after Bosworth, 2005): (a) 31 Ma (Early Oligocene): initiation of magmatism in Afar and in the Arabian plate (Older Harrats and Yemen Trap Series); (b) 24 Ma (Oligocene–Miocene transition) rapid spread of extension throughout Red Sea; (c) 5 Ma (Miocene–Pliocene transition) rift–drift transition in south-central Red Sea; (d) the panel shows the hypothetical layout 10 million years from now.

In the central part of the Red Sea (from ~ 20°N to 24°N) an axial rift valley is absent, and it is characterized by a number of deeps with basaltic activity, separated by inter-trough zones. Further to the north, next to Zabargad Island, the Red Sea axis is displaced laterally by about 100 km along a fracture zone, and a few scattered deeps with basaltic activity have been identified.

The lavas erupted in the Red Sea trough are olivine tholeiites with low K_2O (<1.0%) and low TiO_2 (≤2.0%). The central deeps show the typical features of oceanic basalts (MORB), with low incompatible elements (Rb, Sr, Ba, Nb, La, Sr and Ta), $^{87}Sr/^{86}Sr < 0.703$, and $^{143}Nd/^{144}Nd > 0.513$. However, the chemical and isotope characteristics of the floor basalts show marked changes along the Red Sea. Starting from the central deeps, towards both north and south the basalts display variable increases of incompatible element contents, Sr and Pb isotope ratios, as well as decreases of Nd isotope ratios. Such features support the contribution from an "enriched" mantle component(s). A reversal of the general increase of "enriched" geochemical characteristics towards south occurs at the island of Jabal at-Tair, with the reprise of more primitive features, which then continue to increase further south.

In conclusion, the Red Sea exhibits the characteristics of an ocean in formation and, based on the present day local seismic activity and submarine volcanic activity, it could likely become the sixth ocean on the planet (Figure 15d).

ACKNOWLEDGMENTS

The author is grateful to Orlando Vaselli for its accurate review and comments that greatly helped to improve the manuscript.

This work was financially supported by the Institute of Geosciences and Earth Resources, National Research Council (IGG-CNR), Section of Florence.

REFERENCES

Almalki K.A., Betts P. G., Ailleres L. (2015). The Red Sea – 50 years of geological and geophysical research. *Earth Sci. Rev.* 147, 109-140.

Altherr R., Henjes-Kunst F., Baumann A. (1990). Asthenosphere versus lithosphere as possible sources for basaltic magmas erupted during formation of the Red Sea: constraints from Sr, Pb and Nd isotopes. *Earth. Planet. Sci. Lett.* 96, 269–286.

Altherr R., Henjes-Kunst F., Puchelt H., Bauman A. (1988). Volcanic activity in the Red Sea axial trough—evidence for a large mantle diapir? *Tectonophysics* 150, 121–133.

Baker B. H. (1970). The structural pattern of the Afro-Arabian rift system in relation to plate tectonics. *Philos. Trans. Royal Soc.* London Series A 267, 383–391.

Baker J., Menzies M., Snee L. (1994). Stratigraphy, $^{40}Ar/^{39}Ar$ geochronology and geochemistry of flood volcanism in Yemen. *Mineral. Mag.* 58A, 42–43.

Baker J., Snee L., Menzies M. (1996). A brief Oligocene period of flood volcanism in Yemen: implications for the duration and rate of continental flood volcanism at the Afro-Arabian triple junction. *Earth Planet. Sci. Lett.* 138, 39–55.

Barberi F., Borsi S., Ferrara G., Marinelli G., Santacroce R., Tazieff H., Varet J. (1972). Evolution of the Danakil Depression (Afar, Ethiopia) in light of radiometric age determinations. *J. Geol.* 80, 720–729.

Barrat J. A., Jahn B. M., Joron J. L., Auvray B., Hamdi H. (1990). Mantle heterogeneity in northeastern Africa: evidence from Nd isotopic composition and hygromagmaphyle element geochemistry of basal- tic rocks from the Gulf of Tadjoura and southern Red Sea regions. *Earth Planet. Sci. Lett.* 101, 233–247.

Bayer H-J., Hötzl H., Jado A. R., Roscher B., Voggenreiter W. (1988). Sedimentary and structural evolution of the northwest Arabian Red Sea margin. *Tectonophysics* 153, 137–151.

Bellahsen N., Faccenna C., Funiciello F., Daniel J. M., Jolivet L. (2003). Why did Arabia separate from Africa? Insights from 3-D laboratory experiments. *Earth Planet. Sci. Lett.* 216, 365-381.

Berhe S. M. (1986). Geologic and geochronologic constraints on the evolution of the Red Sea-Gulf of Aden and Afar depression. *J. Afr. Earth Sci.* 5, 101–117.

Bischoff J. L., Manheim F. T. (1969). Economic Potential of the Red Sea Heavy Metal Deposits. In: Degens E. T., Ross D. A. (eds.) *Hot Brines and Recent Heavy Metal Deposits in the Red Sea.* Springer, Berlin, Heidelberg.

Bohannon R. G., Naesar C. W., Schmidt D. L., Zimmerman R. A. (1989). The timing of uplift, volcanism and rifting peripheral to the Red Sea: a case for passive rifting? *J. Geophys. Res.* 94, 1683–1701.

Bonatti E. (1985). Punctiform initiation of seafloor spreading in the Red Sea during transition from a continental to an oceanic rift. *Nature* 316, 33–37.

Bonatti E., Cipriani A., Lupi L. (2015). The Red Sea: Birth of an Ocean. In: Rasul N. M. A., Stewart I. C. F. (eds.) *The Red Sea*, Springer Earth System Sciences, Springer, Berlin, Heidelberg.

Bonatti E., Hamlyn P., Ottonello G. (1981). Upper mantle beneath a young oceanic rift—peridotites from the island of Zabargad (Red- Sea). *Geology* 9, 474–479.

Bonatti E., Ottonello G., Hamlyn P. R. (1986). Peridotites from the island of Zabargad (Red Sea). *J. Geophys. Res.* 91, 599–631.

Bosworth W., Darwish M., Crevello P., Taviani M., Marshak S. (1996). Stratigraphic and structural evolution of Zabargad Island (Red Sea, Egypt) since the early cretaceous. In: El Youssef S. A. (ed.) *Proceedings of the 3rd international conference on geology of the Arab World* 1, 161–190.

Bosworth B., Huchon P., McClay K. (2005). The Red Sea and Gulf of Aden basins. *J. Afr. Earth Sci.* 43, 334–378.

Burke K., Dewey J. F. (1974). Two plates in Africa in the cretaceous? *Nature* 249, 313–316

Camp V. E., Hooper P. R., Roobol M. J., White D. L. (1987). The Madinah eruption, Saudi Arabia: Magma mixing and simultaneous extrusion of three basaltic chemical types. *Bull. Volcanol.* 49, 489-508.

Camp V. E., Roobol M. J., Hooper P. R. (1991). The Arabian continental alkali basalt province: Part II, evolution of Harrats Khaybar, Ithnayn, and Kura, Kingdom of Saudi Arabia. *Geol. Soc. Amer. Bulletin* 103, 363-391.

Chazot G., Bertrand H. (1993). Mantle sources and magma-continental crust interactions during early Red Sea Gulf of Aden rifting in southern Yemen: elemental and Sr, Nd, Pb isotope evidence. *J. Geophys. Res.* 98(B2), 1819–1835.

Chernet T., Hart W. K., Aronson J. L., Walter R. C. (1998). New age constraints on the timing of volcanism and tectonism in the northern Main Ethiopian Rift-southern Afar transition zone (Ethiopia). *J. Volcanol. Geoth. Res.* 80, 267–280.

Cochran J. R. (1983). A model for development of Red Sea. *Am. Assoc. Pet. Geol. Bull.* 67, 41–69.

Coleman R. G. (1993). *Geologic evolution of the Red Sea. Oxford monographs on geology and geophysics*, vol. 24. Oxford University Press, Oxford, 186 pp.

Coleman R. G., McGuire A. V. (1988). Magma systems related to the Red Sea opening. *Tectonophysics* 150, 77–100.

Coulié E., Quidelleur X., Gillot P-Y., Courtillot V., Lefèvre J-C., Chiesa S. (2003). Comparative K-Ar and Ar/Ar dating of Ethiopian and Yemenite Oligocene volcanism: implications for timing and duration of the Ethiopian traps. *Earth. Planet. Sci. Lett.* 206, 477–492.

Courtillot V., Armijo R., Tapponnier P. (1987). Kinematics of the Sinai triple junction and a two phase model of Arabia-Africa rifting. In: Coward M. P., Dewey J. F., Hancock P. L. (eds.) Continental extensional tectonics. *Geol. Soc. London Spec. Publ.* 28, 559–573.

Degens E. T., Ross D. A. (eds.) (1969). *Hot brines and recent heavy metal deposits in the Red Sea a geochemical and geophysical account.* Springer, New York, 571 pp.

Deniel C., Vidal P., Coulon C., Vellutini P. J., Piguet P. (1994). Temporal evolution of mantle sources during continental rifting: the volcanism of Djibouti (Afar). *J. Geophys. Res.* 99, 2853–2869.

Eissen J. P., Juteau T., Joron J. L., Dupré B., Humler E., Al'Mukhamedov A. (1989). Petrology and geochemistry of basalts from the Red Sea axial rift at 18°North. *J. Petrol.* 30(4), 791–839.

Furman T., Bryce J., Rooney T., Hanan B,. Yirgu G., Ayalew D. (2006). Heads and tails: 30 million years of the Afar plume. In: Yirgu G., Ebinger C. J., Maguire P. K. H. (eds.) The Afar Volcanic Province within the East African Rift System. *Geol. Soc. Spec. Publ.* 259, 95–119.

Gass I. G. (1970). The evolution of volcanism in the junction area of the Red Sea, Gulf of Aden and Ethiopian rifts. *Philos. Trans. Royal Soc. London* A 267, 369–382.

George R., Rogers N., Kelley S. (1998). Earliest magmatism in Ethiopia: evidence for two mantle plumes in one flood basalt province. *Geology* 26, 923–926.

Girdler R. W., Southren T. C. (1987). Structure and evolution of the northern Red Sea. *Nature* 330, 716–721.

Haase K. M., Muhe R., Stoffers P. (2000). Magmatism during extension of the lithosphere: geochemical constraints from lavas of the Shaban Deep, northern Red Sea. *Chem. Geol.* 166, 225-239.

Halliday A. H., Lee D. C., Tommasini S., Davies G. R., Paslick C., Fitton J. G., James D. E. (1995). Incompatible trace elements in OIB and MORB and source enrichment in the sub-oceanic mantle. *Earth. Plan. Sci. Lett.* 133, 379–395.

Hart S. R., Hauri E. H., Oschmann L. A., Whitehead J. A. (1992). Mantle plumes and entrainment: isotopic evidence. *Science* 256(5056), 517–520.

Hart W. K., WoldeGabriel G., Walter R. C., Mertzman S. A. (1989). Basaltic volcanism in Ethiopia: constraints on continental rifting and mantle interactions. *J. Geophys. Res.* 94, 7731–7748.

Ilani S., Harlavan Y., Tarawneh K., Rabba I., Weinberger R., Ibrahim K., Peltz S., Steinitz G. (2001). New K-Ar ages of basalts from the Harrat

Ash Shaam volcanic field in Jordan: implications for the span and duration of the upper-mantle upwelling beneath the western Arabian plate. *Geology* 29, 171–174.

Jarrige J. J., Ott d'Estevou P., Burollet P. F., Thiriet J. P., Icart J. C., Richert J. P., Sehans P., Montenat C., Prat P. (1986). Inherited discontinuities and Neogene structure: the Gulf of Suez and northwestern edge of the Red Sea. *Philos. Trans. R. Soc.* London A 317, 129–139.

Joffe S., Garfunkel Z. (1987). Plate kinematics of the circum Red Sea—a re-evaluation. *Tectonophysics* 141, 5–22.

Jónsson S., Xu W. (2015). Volcanic Eruptions in the Southern Red Sea During 2007–2013. In: Rasul N. M. A., Stewart I. C. F. (eds.), *The Red Sea*. Springer Earth System Sciences, Springer, Berlin, Heidelberg.

Joron J. L., Treuil M. (1989). Hygromagmaphile element distributions in oceanic basalts as fingerprints of partial melting and mantle heterogeneities: a specific approach and proposal of an identification and modelling method. In: Saunders A. D., Norry M. J. (Eds.), Magmatism in the Ocean Basins, *Geol. Soc. Spec. Publ.* 42, 277–299.

Keir D., Bastow I. D., Pagli C., Chambers E. L. (2012). The development of extension and magmatism in the Red Sea rift of Afar. *Tectonophysics* 607, 98-114.

Kenea N. H., Ebinger C. J., Rex D. C. (2001). Late Oligocene volcanism and extension in the southern Red Sea Hills, Sudan. *J. Geol. Soc. London* 158, 285–294.

Khan M. A. (1975). The Afro-Arabian rift system. *Science Progress* 62, 207–236.

Le Pichon X., Francheteau J. (1978). A plate tectonic analysis of the Red Sea—Gulf of Aden area. *Tectonophysics* 46, 369–406.

Ligi M., Bonatti E., Bortoluzzi G., Cipriani A., Cocchi L., Caratori Tontini, F., Carminati E., Ottolini L., Schettino A., (2012). Birth of an ocean in the Red Sea: initial pangs. *Geochem. Geophys. Geosyst.* 13, Paper Q08009.

Marty B., Appora I., Barrat J A., Deniel C., Vellutini P., Vidal P. (1993). He, Ar, Sr, Nd, and Pb isotopes in volcanic rocks from afar: evidence for a primitive mantle component and constraints on magmatic sources. *Geochemistry* 27:223–232.

Marty B., Pick R., Gezahegn Y. (1996). Helium isotopic variations in Ethiopian plume lavas: nature of magmatic sources and limit on lower mantle contribution. *Earth. Planet. Sci. Lett.* 144, 223–237.

Mattash M. A (2008). Younger volcanic fields of Yemen with focus on Jabal at-Tair active volcano in the Red Sea. In: *Int. Geol. Cong. Abstracts* 33, Abstract 1287199.

Mattash M. A., Pinarelli L., Vaselli O., Minissale A., Al-Kadasi M., Shawki M. N., Tassi F. (2013). Continental Flood Basalts and Rifting: Geochemistry of Cenozoic Yemen Volcanic Province. *Int. J. Geosci.* 4(10), 1459-1466.

Mattash M. A., Pinarelli L., Vaselli O., Minissale A., Jaimes-Viera M. C., Al-Kadasi M., Shawki M. N., Tassi F. (2014). Geochemical evolution of southern Red Sea and Yemen flood volcanism: evidence for mantle heterogeneity. *Arab. J. Geosci.* 7(11), 4831-4850.

McClusky S., Reilinger R., Ogubazghi G., Amleson A., Haleb B., Vernant P., Sholan J., Fisseha S., Asfaw L., Bendick R., Kogan L. (2010). Kinematics of the southern Red Sea–Afar Triple Junction and implications for plate dynamics. *Geoph. Res. Lett.* 34, L05301.

McKenzie D. P., Davies D., Molnar P. (1970). Plate tectonics of the Red Sea and East Africa. *Nature* 226, 243–248.

Mitchell N. C., Park Y. (2014). Nature of crust in the central Red Sea. *Tectonophysics* 628. 123-139.

Moretti I., Chénet P. Y. (1987). The evolution of the Suez rift: a combination of stretching and secondary convection. *Tectonophysics* 133, 229–234.

Moufti M. R. H., Nemeth K. (2016). Harrat Rahat: The Geoheritage Value of the Youngest Long-Lived Volcanic Field in the Kingdom of Saudi Arabia. In: Moufti M. R. H. and Nemeth K (eds.), *Geoheritage of Volcanic Harrats in Saudi Arabia*. Springer, pp. 33-120.

Moufti M. R. H., Nemeth K., Lindsay J. M. (2013). The 1256 AD Al Madinah historic eruption geosite as the youngest volcanic chain in the Kingdom of Saudi Arabia. *Int. J. Earth Sci.* 102, 1069-1070.

Orihashi Y., Nagao K., Al-Jailani A., Hanan B. B. (2001). Primordial helium isotope signature from Plio-Quaternary alkaline basalts in Yemen. *Island Arc* 10, 145–157.

Pik R., Deniel C., Coulon C., Yirgu G., Marty B. (1999). Isotopic and trace element signatures of Ethiopian flood basalts: evidence for plume–lithosphere interactions. *Geochim. Cosmochim. Acta* 63, 2263–2279.

Pinarelli L., Mattash M. A., Vaselli O., Minissale A., Tassi F. (2015): Geochemical evolution of mantle sources during continental rifting: the volcanism of the Red Sea. *Rend. Online Soc. Geol. It.* vol. 35(2), p. 53.

Prodehl C., Fuchs K., Mechie J. (1997). Seismic-refraction studies of the Afro-Arabian rift system—a brief review. *Tectonophysics* 278, 1–13.

Rochette P., Tamrat E., Féraud G., Pik R., Courtillot V., Kefeto E., Coulon C., Hoffmann C., Vandamme D., Yirgu E. (1997). Magnetostratigraphy and timing of the Oligocene Ethiopian traps. *Earth. Planet. Sci. Lett.* 164, 497–510.

Rogers N. W. (1993). The isotope and trace element geochemistry of basalts from the volcanic islands of the southern Red Sea. In: Prichard H. M., Alabaster T., Haris N. B., Neary C. R. (eds.) Magmatic processes and plate tectonics. *Geol. Soc. London. Spec. Publ.* 76, 455–467.

Rudnick R. L., Gao S. (2003). Composition of the continental crust. *Treatise Geochem.* 3, 1–64.

Schilling J-G., Kingsley R. H., Hanan B. B., McCully B. L. (1992). Nd–Sr–Pb isotopic variations along the Gulf of Aden: evidence for mantle plume continental lithosphere. *J. Geophys. Res.* 97, 10927–10966.

Sebai A., Zumbo V., Féraud G., Bertrand H., Hussain A. G., Giannerini G., Campredon R. (1991). $^{40}Ar/^{39}Ar$ dating of alkaline and tholeiitic magmatism of Saudi Arabia related to the early Red Sea rifting. *Earth Planet. Sci. Lett.* 104, 473–487.

Stracke A., Bizimis M., Salters V. J. M. (2003). Recycling oceanic crust: Quantitative constraints. *Geochem. Geophys. Geosyst.*, 4 (3), Q8003.

Stracke A., Hofmann A. W. (2005). FOZO, HIMU, and the rest of the mantle zoo. *Geochem. Geophys. Geosyst.* 6, Q05007.

Steckler M. S. (1985). Uplift and extension at the Gulf of Suez: indications of induced mantle convection. *Nature* 317, 135–139.

Sun S. S., McDonough W. F. (1989). Chemical and isotopic systematics of oceanic basalts: implications for mantle composition and processes. In: Saunders A. D., Norry M. J. (eds.) Magmatism in the ocean basins. *Geol. Soc. London Spec. Publ.* 42, 313–345.

Szymanski E. (2013). *Timing, kinematics and spatial distribution of Miocene extension in the central Arabian margin of the Red Sea rift system.* PhD thesis, University of Kansas, Lawrence, 430 pp.

Szymanski E., Stockli D. F., Johnson P. R., Hager C. (2016). Thermochronometric evidence for diffuse extension and two-phase rifting within the Central Arabian Margin of the Red Sea Rift. *Tectonics* 35, 2863-2895.

Ukstins I. A., Renne P. R., Wolfenden E., Baker J., Ayalew D., Menzies M. (2002). Matching conjugate volcanic rifted margins: $^{40}Ar/^{39}Ar$ chrono-stratigraphy of pre- and syn-rift bimodal flood volcanism in Ethiopia and Yemen. *Earth Planet. Sci. Lett.* 198, 289–306.

Varet J. (1978). *Geology of central and southern Afar (Ethiopia and Djibouti Republic).* Edition CNRS, Paris, 118 pp.

Vellutini P. (1990). The Manda-Inakir Rift, Republic of Djibouti: a comparison with the Asal Rift and its geodynamic interpretation. *Tectonophysics* 172, 141–153.

Volker F., Altherr R., Jochum K. P., McCulloch M. T. (1997). Quaternary volcanic activity of the southern Red Sea: new data and assessment of models on magma sources and afar plume–lithosphere interaction. *Tectonophysics* 278, 15–29.

Volker F., McCulloch M. T., Alther R. (1993). Submarine basalts from the Red Sea: new Pb, Sr and Nd isotopic data. *Geophys. Res. Lett.* 20(10), 927–930.

White W. M. (1985). Sources of oceanic basalts: Radiogenic isotopic evidence. *Geology* 13 (2), 115-118.

Whitmarsh R. B, Weser O. E.; Ross D. A., Syed A., Boudreaux J. E., Coleman R. G., Fleisher R. L., Girdler R. W., Jipa D., Kidd R. B., Mallik T. K., Manheim F. T., Matter A., Nigrini C. A., Siddiquie H. N., Stoffers P., Supko P. R. (1974). Chemical determinations of iron-manganese rich sediments, Sites 221, DSDP Leg 23. In: Whitmarsh R. B., Weser O. E., Ross D. A. (eds.) *Initial Reports of the Deep Sea Drilling Project*, U. S. Government Printing Office XXIII, 167-210.

Willbold M., Stracke A. (2006). Trace element composition of mantle end-members: implications for recycling of oceanic and upper and lower continental crust. *Geochem. Geophys. Geosyst.* 7, Q04004.

Xu W., Jónsson S. (2014). The 2007–8 volcanic eruption on Jebel at Tair island (Red Sea) observed by satellite radar and optical images. *Bull. Volcanol.* 76, 795.

Zanettin B., Justin-Visentin E., Piccirillo E. M. (1978). Volcanic succession, tectonics and magmatology in central Ethiopia. *Atti Mem. Accad. Padovana Sci. Lett. Arti* 90, 5–19.

Zeyen H., Volker F., Wehrle V., Fuchs K., Sobolev S. V., Altherr R. (1997). Styles of continental rifting: crust–mantle detachment and mantle plumes. *Tectonophysics* 278, 329–352.

Zindler A., Hart S. (1986). Chemical Geodynamics. *Annual Rev. Earth Planet Sci.* 14, 493-571.

Zumbo V., Féraud G., Bertrand H., Chazot G. (1995). $^{40}Ar/^{39}Ar$ chronology of Tertiary magmatic activity in southern Yemen during the early Red Sea-Aden rifting. *J. Volcanol. Geoth. Res.* 65, 265–279.

Reviewed by: Orlando Vaselli, Department of Earth Sciences, University of Florence, Florence, Italy.

BIOGRAPHICAL SKETCH

Laura Pinarelli, PhD

Affiliation: Institute of Geosciences and Earth Resources (IGG), Section of Florence, National Research Council (CNR)

Office Address: 4 Via G. La Pira, 50121 Firenze (Italy)

Email: lapina@igg.cnr.it

Web Page: http://www.cnr.it/people/laura.pinarelli

Education:

1987 - PhD in Earth Sciences (Thesis in Geochemistry), University of Florence (Italy)

1982 - Master Degree in Earth Sciences (Thesis in Geochemistry), University of Florence (Italy)

Professional Appointments:

2006-present: Senior Researcher at IGG – CNR, Florence (Italy)

2000-2005: Researcher at IGG – CNR, Florence (Italy

1998-2000: Researcher at Centre for Study of Minerogenesis and Applied Geochemistry – CNR, Florence (Italy)

1995: Visiting Scientist at Laboratory for Isotope Geology of Max Plank Institute für Chemie, Mainz (Germany)

1987: Visiting Scientist at Laboratory for Isotope Geology and Masspectrometry of the Institute for Crystallography and Petrography – ETH, Zürig (Switzerland)

1986-1997: Researcher at Institute of Geochronology and Isotope Geochemistry – CNR, Pisa (Italy)

Research Experience:

- Isotope Geochemistry and Petrology: Relationships between magmatism and geodynamics: Aeolian Arc, Plio-quaternary province of Italy, Aegean Arc, Red Sea rift., Cenozoic volcanism of Alps (Italy), Greece, Bulgaria, Yemen. Characterization of Palaeozoic basements: Southern Alps, Calabria (Italy), and Spain.
- Geochronology: Rb-Sr, U-Pb and Pb-Pb.
- Cultural Heritage: Pb isotope characterization and provenance study of metallic manufacts.

Professional Experience:

- Person in charge of Research Projects financed by: Department of Civil Protection, Italian Ministry of the Cultural Heritage, CNR.
- Participant in Research Projects financed by: European Economic Community (EEC), Italian Ministry of Education, University and Research (MIUR), and CNR Agency.
- Member of the Board of Overseers of IGG.
- Referent of IGG for transparency and integrity.
- Member of the Editorial Board and Scientific Committee of the journal "Scienze e Ricerche," ISSN: 2283-5873 since 2016.
- Member of the committee for updating the IGG website, 2015-2016.
- Member of the committee for assignment of a CNR fellowship, Dec. 2015.

- Member of the Jury of National Prize of Scientific Dissemination, sponsored by CNR and AIRI (Associazione Italiana Ricerca Industriale). Editions of 2015 and 2016.
- Appraiser of research projects for the University of Insubria, 2016-2017.
- Expert for the Research & Innovation Portal of the European Commission since 2015.
- Member of the Jury for the Innovation Award assigned by CNR. Edition 2015.
- Member of the committee for assignment of a CNR research grant, Jan. 2015.
- Appraiser of "Futuro in Ricerca" Projects of MIUR (Italian Ministry of University and Research) since 2013.
- Member of Register of peer reviewers of MIUR for the evaluation of research programs (PRIN) and research products (VQR) since 2012.
- Member of Register of Accredited Experts in Industrial Research of MIUR since 2012.
- Member of Scientific Committee of the Conference Geoitalia 2011, Torino, Sept. 19-23, 2011.
- Member of the Jury of the Ezio Tongiorgi prize in Geochemistry assigned by Società Geochimica Italiana (SoGeI), edition 2009.
- Member of the Assembly of Geoitalia as delegate of SoGeI since 2007.
- Guest Scientist at Max Plank Institute für Chemie - Mainz (Germany), Feb.-June, 1995.
- Guest Scientist at ETH - Institute for Isotope Geology and Mineral Resources - Zurich (Switzerland), Sept.-Dec., 1987.

Membership to Associations:

- Member of Associazione Italiana del Libro since 2013.
- Member of Società Geologica Italiana (SGI) since 2010.
- Founding Member of Società Geochimica Italiana (SoGeI) since 1999.

- Member of Federazione Italiana Scienze della Terra (FIST) since 1996.
- Member of Società Italiana di Mineralogia e Petrologia (SIMP*)* from 1982 to 1992.

Publications in the Last 3 Years:

1) Pinarelli L., Christofides G., Pipera K., Soldatos T., Koroneos A., Vaselli O. (2016): K-rich magmatism in extensional settings: the Oligocene lower Miocene Evros volcanic rocks from north eastern Greece (Thrace*). 88° National Congress of the Italian Geological Society* "Geosciences on a changing planet: learning from the past, exploring the future," Napoli (Italy) 7-9 September. *Rend. Online Soc. Geol. It.* Suppl. 12 vol. 40, p. 194. [DOI: 10.3301/ROL.2016.79].
2) Pinarelli L., Vaselli O., Capaccioni B., Giocada A., Downes H. (2016): The Spiaggia Lunga mafic enclaves (Vulcano, Aeolian Islands): testimony of an older reservoir. *88° National Congress of the Italian Geological Society* "Geosciences on a changing planet: learning from the past, exploring the future," Napoli (Italy), 7-9 September. *Rend. Online Soc. Geol. It.* Suppl. 12 vol. 40, p. 259. [DOI: 10.3301/ROL.2016.79].
3) Pinarelli L., Christofides G., Pipera K., Soldatos T., Koroneos A., Pecksay Z.. (2015): Mantle heterogeneity and crustal contamination in Tertiary extensive magmatism in Southern Rhodopes: geochemical and isotope evidence from Evros volcanic rocks (Greece). Joint Conference SIMP-AIV-SoGeI-SGI "Il Pianeta Dinamico: sviluppi e prospettive a 100 anni da Wegener," Florence (Italy), 2-4 September 2015. *Rend. Online Soc. Geol. It.* Suppl. 2 vol. 35, p. 96. [DOI: 10.3301/ROL.2015.131].
4) Pinarelli L., Mattash M.A., Vaselli O., Minissale A., Tassi F. (2015): Geochemical evolution of mantle sources during continental rifting: the volcanism of the Red Sea. *Joint Conference SIMP-AIV-SoGeI-SGI* "Il Pianeta Dinamico: sviluppi e prospettive a 100 anni da Wegener," Florence (Italy), 2-4 September. *Rend. Online Soc. Geol. It.* Suppl. 2 vol. 35, p. 53. [DOI: 10.3301/ROL.2015.131].

5) Christofides G., Pinarelli L., Pipera K., Soldatos T., Koroneos A., Pecskay Z. (2015): Geochemical features of the depression basin-related magmatism in Southeastern Rhodope Massif and Circum-Rhodope Belt: the Tertiary Evros volcanic rocks (Western Thrace, Northeastern Greece). *Acta Vulcanol.* 25(1-2), 43-56. [DOI: 10.3301/ROL.2015.131].
6) Mattash M.A., Pinarelli L., Vaselli O., Minissale A., Jaimes-Viera M.C., Al-Kadasi M., Shawki M. N., Tassi F. (2014): Geochemical evolution of southern Red Sea and Yemen flood volcanism: evidence for mantle heterogeneity. *Arab. J. Geosci.* 7(11), 4831-4850. [DOI:10.1007/s12517-013-1120-1].
7) Pinarelli L., Nisi B., Vaselli O., Minissale A., Buccianti A., Tassi F. (2014): Major, trace element, and Sr isotope geochemistry of surface and ground waters in the Chiavenna Valley (Sondrio, Northern Italy). *Rend. Online Soc. Geol. It.* 30, 62-65. [DOI: 10.3301/ROL.2014.13].

In: Red Sea
Editor: Delbert Robbins

ISBN: 978-1-53613-200-7

Chapter 2

MINERAL RESOURCES IN THE RED SEA/GULF OF ADEN: POLYMETALLIC-BARITE MINERALIZATIONS AT WADI MASILAH BASIN (YEMEN)

Laura Pinarelli, PhD[1,*] and Mohamed A. Mattash[2]

[1]Institute of Geosciences and Earth Resources, Section of Florence, National Research Council (CNR), Florence, Italy
[2]Ministry of Oil and Mineral Resources, Geological Survey and Minerals Resources Board, Sana'a, Yemen

ABSTRACT

Yemen is characterized by a considerable number of metallic and non-metallic raw materials, including base metals, precious metals, and dimension stones. The Wadi Masilah Basin, in the southeasternmost part of Yemen, hosts very promising base metal and barite prospects. This

---

* Corresponding Author Email: lapina@igg.cnr.it.

terrain has undergone a complex geological and structural history, including oblique divergence (transtension) related to the opening phases of the greater Red Sea – Gulf of Aden rift system, with NW-SE and NE-SW major fault zones.

Barite occurs either singly or in association with galena, hemimorphite, willemite, zincite, smithsonite, descloizite, celestine, quartz, calcite, dolomite, iron and manganese oxides, and gypsum, connected mainly to the Cretaceous carbonates (Saa'r Formation). The mineralisations crop out as either vein fillings or stratiform type. $BaSO_4$ contents between 95 and 98% and specific gravity values up to 4.37 g/cm^3 are indicative of high quality ores.

Sulfide mineralisations consist of sphalerite, galena and pyrite. They occur in association with barite veins or form fracture-filling in limestones and silicified dolostones. Available data indicate several lead-zinc ore bodies at various depths, some of which are considerably large with dimensions attaining 60 m x 30 m x 50 m, and tonnages reaching thousands of tons.

Nonsulfide minerals comprise hemimorphite, willemite, smithsonite, hydrozincite, and zincite, as well as cerussite, hydrocerussite, anglesite, wulfenite, and rarely pyromorphite. Zinc contents in the mineralised veins range from low (1-8 wt.%) to high grades (30-50 wt.%). The above mentioned minerals occur in a variety of structures, such as: shear zones, fractures, vugs, cavities, brecciated veins, and open spaces of the vein barite sidewalls, and connected mainly to the dolostone sequence.

Vanadium ores mostly occur as descloizite and rarely vanadinite, in association with barite, calcitized veins and brecciated zones, and dolomitic limestones, with vanadium contents up to 12.30 wt.%. These prospects contain one of the most remarkable occurrences of descloizite known, and besides lead-zinc, descloizite is probably the mineral resource for which the Wadi Masilah Basin is most promising.

Mn and Fe ores are also present, whose mineralogy consists of hollandite, romanechite, cryptomelane, pyrolusite, hematite, goethite and limonite. They include several discordant irregular veins. The results of geochemical analyses reveal high quality Mn ores, with an average MnO content of 64%.

The orebodies in all prospects consist mainly of fracture-filling in limestones, dolostones, and silicified limestones. They originated from both hydrothermal processes and the action of slightly acidic groundwater. The prospecting area is structurally controlled and located in a NW-SE major fault zone and its parallel lineaments. Tectonic uplift and intensive fracturing systems, along with favorable climatic conditions, gave rise to the development of supergene Zn, Pb, V, Mn and Fe mineralisations.

Keywords: Yemen, Wadi Masilah Basin, polymetallic mineralisations, barite

1. Introduction

Non-renewable raw materials are fundamental for industrialised economies. In spite of the rapid growth of renewable energy, industrialised countries still rely heavily on non-renewable resources. Moreover, many emerging economies are now entering the phase of strong industrialisation that requires significant amounts of energy, mineral resources, metals and other raw materials. Mineral deposits and metals are among the most important materials required by almost all countries for construction, production of tools and equipment, for wiring and manufacturing of automobile body parts. Even though mineral deposits are comprised mostly of naturally occurring elements, their extraction and processing is very expensive. Nowadays, there is an increasing interest in the supergene polymetallic deposits, particularly after successfully developing of a number of procedures that facilitate their processing and concentration (e.g., Falagàn et al., 2017).

Yemen is characterized by remarkable deposits of metallic and non-metallic raw materials, including base metals, precious metals, and dimension stones. A variety of ancient mines were found across the country, of gold, silver, copper, iron, lead, etc., which were exploited by the ancient Yemenis to extract their metals and oxides, used to produce metal objects, cosmetics, pigments, ceramics, ornaments. The last Century has witnessed several intensive explorations, including geological, geochemical, and geophysical surveys, and thematic mapping. Such activities were systematically undertaken by Yemeni, European, and North American Institutions and Societies. Recently, a new Mining Law has been introduced to encourage foreign and local companies to invest in the minerals, petroleum, and gas sectors, as well as in the geothermal energy resources.

This chapter reports a synthesis of many years of geological and geochemical investigations on the Wadi Masilah Basin, in particular at Alkohl1, Alkohl2, and Alqalanah areas, which contain the first barite occurrences in Yemen, as well as supergene non-sulfide zinc minerals, and vanadium rich zones (Mattash et al., 2007a and b; 2008; 2009; 2010). These outcrops are amongst the metallic-nonmetallic occurrences of Yemen that are already partially exploited.

2. Geological Background

Yemen is situated in the southern part of the Arabian Peninsula and its basement complex is part of the Arabian Shield (Figure 1). The geological evolution of Yemen was driven by the plate motions that broke Pangea apart in the Mesozoic and formed the greater Red Sea–Gulf of Aden rift system. The stratigraphy and regional geology of Yemen was established by detailed work of Beydoun (1964), Powers et al., (1966), Beydoun and Greenwood (1968), Hughes and Clarke (1988), Haitham and Nani (1990), Paul (1990), Bott et al., (1992), Hughes and Beydoun (1992), Schlumberger (1992), and Beydoun et al., (1993). Hydrocarbon exploration activity after the '90s allowed for a revised synthesis of basin evolution in Yemen, such as the work of Redfern and Jones (1995), Ellis et al., (1996), and Beydoun et al., (1996, 1998).

The Wadi Masilah Basin (WMB), in the southeasternmost part of Yemen, is one of the twelve sedimentary basins identified in Yemen (Figure 1), categorised into three groups based on their age: Palaeozoic, Mesozoic and Cenozoic (As-Saruri and Sorkhabi, 2016). The WMB, developed during the fragmentation of Gondwana in Mesozoic times, is elongated with orientation sub-parallel to the Red Sea rift, and is filled with syn- and post-rift sediments. The succession is characterized by Mesozoic marine sediments, belonging to the Saa'r, Qishn, Fartaq, and Um Er-Rhaduma Formations, overlaid by the Cenozoic Shihr Group deposits. A very thin layer of Quaternary sediments occurs. Several small areas of the WMB are covered by Quaternary basaltic lavas associated to the post-rift phases of the Red Sea and Gulf of Aden (Mattash, 1994, Mattash et al., 2013; 2014).

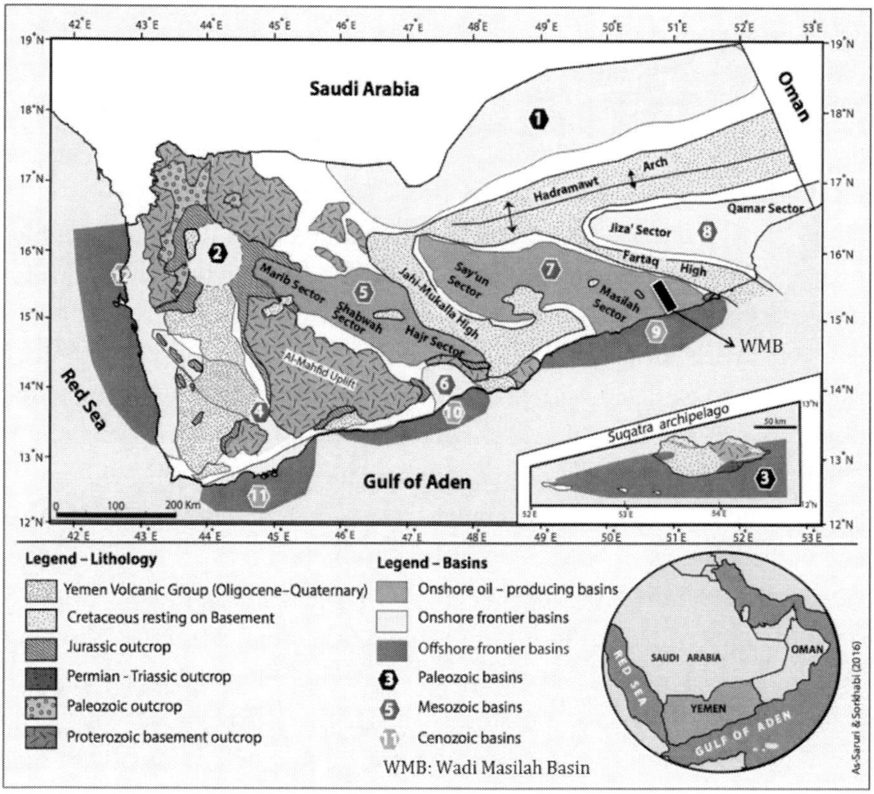

Figure 1. Map of the outcropping geology, major structural highs, and sedimentary basins of Yemen, and location of the study area (WMB).

The WMB has undergone a complex geological and structural history, including oblique divergence (transtension) related to the opening phases of the Red Sea and Gulf of Aden rifts, with NW-SE and NE-SW major fault zones, respectively. As a result, the prospect area has been strongly faulted. Normal as well as thrust faults are characteristic features, whereas fissures and joints are common structural patterns. Small horsts and grabens are locally developed, indicating complex faulting activity. Some parts of the sedimentary sequence are slightly subjected to folding, as well.

3. Occurrence, Mineralogy, and Geochemistry

The common paragenesis in the WMB mineralised prospects is barite, sphalerite, almost entirely oxidized to hemimorphite, willemite, zincite, hydrozincite, smithsonite, galena, partially decomposed to cerussite, hydrocerussite, anglesite, wulfenite, and rare pyromorphite. Calcite is a widespread gangue mineral both in lately recrystallized form, associated to the contacts of barite veins, and as part of the nonsulfide massive phase. In addition, descloizite and less commonly vanadinite also occur. Mn and Fe oxides and hydroxides are common and include hollandite, romanechite, and pyrolusite, as well as hematite, limonite, and goethite, respectively, where hematite is locally forming an important ore mineral. The ore bodies are extensively oxidized and enriched in nonsulfide Zn and Pb minerals and breccias. In addition, chalcedony and jasper occur as latest generation, together with fine-grained transparent hemimorphite crystals. Dolomite, aragonite, siderite, goethite, limonite, gypsum, and clay minerals also occur. Euhedral coarse, medium and fine-grained quartz crystals occur as cavity or druse filling. Paragenetically, it appears that willemite formed first, with zincite, hydrozincite, smithsonite, and hemimorphite following in determined order. However, hydrozincite, smithsonite and hemimorphite may have been emplaced at the same time. There is also evidence of a later generation of zincite. The area was affected by hydrothermal activity that produced dolomitisation, brecciation, limonitisation (after pyrite and probably other sulphides), and intensive silicification in some layers.

The mineralised bodies are brecciated, massive bodies that extend near horizontally in association or in close proximity to the barite veins and throughout the limestones, dolostones, and silicified limestones. They are also exposed at the surface both as fracture filling and dissemination within the silicified dolomitic limestones. The mineralisations are oxidized and diffusely enriched in nonsulfide Zn and Pb minerals and breccia. The extent of oxidation and upgrading of the ores depends on the intensity of circulation of meteoric waters in a relatively young fracture system. WNW-ESE trending lenses of barite-filled fractures are steeply dipping to near vertical.

Two main types of mineralisations can be distinguished based on mineralogy: Alkohl-Kilmy-Alqalanah barite-lead-zinc-iron-manganese-vanadium mineralisation, and Al-Kohl West zinc-lead-barite. The former is associated with the barite veins and their sidewalls (along fractures), whereas the latter is associated with brecciated zones, indicating that they were formed due to extension to compression structural patterns displayed throughout the WMB. The spatial distribution of the WMB mineralised prospects is stratigraphically and structurally controlled.

3.1. Barite

Barite occurs singly or in association with galena, hemimorphite, willemite, zincite, smithsonite, cerussite, descloizite, celestine, quartz, calcite, dolomite, iron and manganese oxides, and less commonly gypsum, connected mainly to the Cretaceous carbonates (Saa'r Formation). Barite deposits at the WMB prospects are considered as medium-high grade, with a probable resource exceeding one million tons.

Barite is mostly characterized by rosette texture (Figure 2a), and has four main types of occurrence: veins and cavity fillings, bedded (stratiform), in association with silicified and breccia zones, and residual.

3.1.1. Veins and Cavity Filling

Barite veins are characterised by either simple or complex mineralisations. The former contains only barite and galena; the latter includes barite, zincite, hemimorphite, willemite, smithsonite, zincite, hydrozincite, descloizite, calcite, as well as Mn and Fe oxides and hydroxides. Barite is frequently found in association with galena, cerussite and hydrocerussite. The grain size generally varies from fine to very fine. The mineralisation within the sidewalls of barite veins may be continuous or brecciated. The veins were formed by the effect of hydrothermal fluids generally deposited in WNW trending faults, fractures, joints, as well as in planes of bedded limestones and dolostones. They are characterized by sharp contacts, extreme variations in length, which exceeds in some cases

1 km, and width, which ranges from 0.6 to 5 m. The depth of the veins is also variable from one place to another, and drilling activity indicates depths up to 6 m, with remnants at a depth of 64 m. In addition, traces of barite were reported from a depth of 109 meters. Since most veins occur near the surface, barite can be easily extracted. Indeed, an attempt to produce barite started in 2010, but the production was ceased due to a difficult political situation affecting the country.

Figure 2. (a) Typical rosette texture of barite. (b) Barite rich breccia with Zn oxides.

Geochemical survey revealed $BaSO_4$ contents between 95 and 98% (Table 1) and specific gravity values up to 4.37 g/cm^3 for most portions of the vein massive material. Such features are indicative of high quality ore in the API international quality classification (Table 2).

Table 1. Chemical composition of barite raw material at Alkohl1

Type	**V**	**V**	**V**	**V**	**V**	**V**	**V**	**V**
Comp. wt.%	MAM-318	MAM-319	B1	B2	B3	B4	B5	B6
SiO_2	0.16	0.41	0.14	0.13	bdl	0.96	bdl	bdl
Al_2O_3	0.12	0.18	bdl	bdl	bdl	bdl	bdl	bdl
Fe_2O_3	0.09	0.07	0.08	0.08	bdl	0.02	bdl	bdl
CaO	0.06	0.18	0.06	0.02	0.03	0.05	0.36	0.07
Na_2O	-	-	0.49	0.47	0.44	0.49	0.49	0.48
MgO	0.1	0.15	bdl	bdl	bdl	bdl	bdl	bdl
MnO	0.04	0.04	bdl	bdl	bdl	bdl	bdl	bdl
SrO	1.85	1.23	-	-	-	-	-	-
L.O.I.	0.18	0.36	0.28	0.38	0.36	0.26	0.46	0.16
$BaSO_4$	97.72	97.33	94.81	95.30	95.14	96.33	98.4	96.60

Type	**B**	**B**	**S**	**S**	**S**
Comp. wt.%	MAM-444	MAM-445	MAM-292	MAM-306	MAM-310
SiO_2	2.57	1.28	65.28	58.69	34.57
Al_2O_3	0.36	0.19	1.82	1.01	0.41
Fe_2O_3	1.3	0.4	4.33	1.52	0.83
CaO	10.69	7.79	0.55	0.47	0.46
Na_2O	-	-	-	-	-
MgO	2.16	0.6	0.35	0.21	0.19
MnO	0.13	0.06	0.03	0.02	0.06
SrO	0.62	0.77	0.26	0.24	0.55
L.O.I.	10.9	6.54	3.42	1.71	1.26
$BaSO_4$	69.60	78.1	24.59	36.43	59.49

V = Veins. B = Bedded; S = assoc. with Silicified breccia. bdl = below detection limit (Al_2O_3 = 0.05%; Fe_2O_3, MgO = 0.01%; MnO = 0.001%). - = not analysed.

3.1.2. Bedded

In the upper part of the sedimentary sequence, bedded occurrences are main constituents. Stratiform barite material is frequently associated with quartz, calcite and, to a lesser extent, descloizite. Barite is fine to very fine-grained, sometimes dark grey in colour (black bedded barite). The beds of barite range in thickness from 0.20 to 0.40 m or more, and extend over an area exceeding 1 km^2. Their $BaSO_4$ contents range from 44 to 70% and in some cases may reach 78% (Table 1). Such lower contents with respect to the vein occurrences are due to the involvement of fine grained associated material, consisting of carbonate, silica, and less commonly iron and manganese compounds and clay, which in some cases may reach 85% of the raw material.

Table 2. API international quality classification of barite ores

Quality		
High	Medium	Low
BaSO4 > 85%	BaSO4=60-84%	BaSO4 < 60%
Density: 4.20-4.37 g/cm^3		

3.1.3. Association with Silicified and Breccia Zones

Such barite mineralisations (Figure 2b) occur either as patches formed on the surface, or as main constituent of the silicified dolomitic limestones. The grade of barite in the silicified dolostones varies considerably from one zone to another, depending upon the activity of hydrothermal solutions and the structure of individual zones. $BaSO_4$ contents range from 25%, to 60% (Table 1).

3.1.4. Residual Barite

Residual occurrences of barite can be found on the surface. They are due to either weathering processes of the underlying dolomitic limestones, or breakdown of barite veins and bedded barite layers. The residual barite ranges from fine to very fine-grained (in the wadi sediments), to lumps and blocks (near the sedimentary beds), weighing from a few kilograms up to dozens of kilograms, respectively. The thickness of such occurrences may attain several meters. The grade of residual barite varies considerably from one place to another, but in some cases it can reach 100 to 500 kilograms per cubic meter.

3.2. Hypogene Sulphide Mineralisations

Hypogene sulphide mineralisations mainly consist of sphalerite, galena and pyrite. Most of the ore bodies are composed of hydrothermally brecciated veins and cavity filling material, settled in deformed limestone and dolostone strata of Jurassic and Cretaceous ages, and possibly connected to a stratabound structure. The possible productive zones for Pb-Zn and other metallic ores in this district are mainly distributed throughout the Alkohl-Wadi Kilmy and Alqalanah areas. The ore bodies mainly follow a NW-SE extensional fault system sub-parallel to the Red Sea extensional zone. Available data indicate several lead-zinc ore bodies at variable depths, ranging between 300 and 500 m, with dimensions attaining 60 m x 30 m x 50 m, and tonnages reaching thousands of tons.

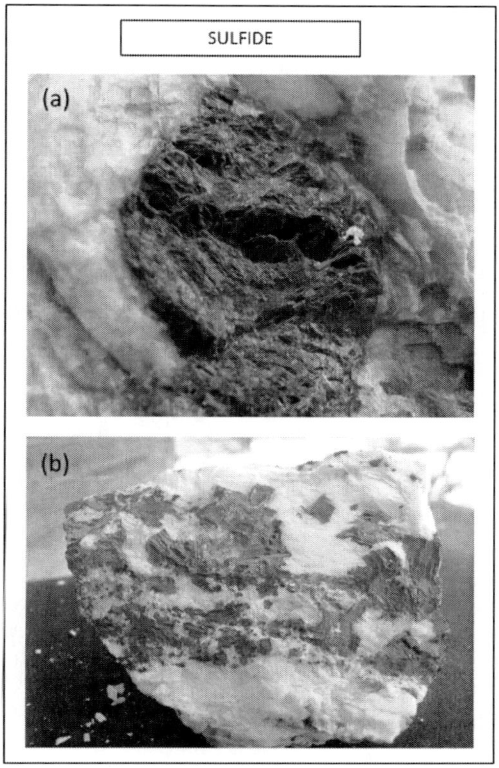

Figure 3. (a) Euhedral coarse grained sphalerite. (b) Galena-rich barite.

3.2.1. Sphalerite

Finer grains of sphalerite are yellow, brown and reddish, whereas coarser ones are mostly black (Figure 3a). Some sphalerite grains are oxidized along their cleavage lines, others are completely altered. Perfect sphalerite crystals have a very distinct adamantine lustre, the recrystallized grains in particular, whereas those subjected to oxidation are resinous and mostly characterized by greenish dark to black colour. The colour variety of sphalerite crystals indicates that Mn, Fe, and Cd isomorphously substitute Zn. The grain size is variable, ranging from microscopic up to 5 cm. Occurrence of chalcopyrite blebs closer to the sphalerite boundaries is an indicator of high temperature of formation within hydrothermal veins.

3.2.2. Galena

Massive galena is commonly associated with barite veins (Figure 3b), sometimes with limestones, dolostones, and silicified zones and veins. It is also found in association with sphalerite and nonsulfide Zn minerals. Galena seems to be a late-formed mineral in the paragenesis, particularly if compared to sphalerite and other associated hypogene sulphides.

Galena is lead-grey, bluish-grey, dark-grey, and sometimes tarnished due to oxidation and weathering. It mainly occurs in cubic (NaCl type) crystals, and its grain size is variable from millimetres to several centimetres. Twinning is present in well crystallized coarse grains. In some cases, recrystallization of galena is developed by natural weathering, forming a granoblastic texture. Skeletal growth was also observed, in which anglesite and wulfenite are developed. Galena is mainly unaltered except for coatings of cerussite, anglesite, less commonly wulfenite, and occasional pyromorphite and plumbonacrite, which are found in minute quantities at a few places, particularly at the Alkohl prospect. Silver content, found as fine-grained mechanical mixture, can reach 300 ppm or more, and averaged 20 ppm throughout the surface galena-rich material. Selenium is also reported with contents up to 108 ppm, whereas Sb and Zn contents may exceed 0.1%.

3.2.3. Pyrite

Pyrite is characterized by metallic lustre and yellow to slightly greenish yellow colours. Crystals are mostly simple cubes or sometimes pentagonal dodecahedra, often occurring in disseminated granular form, in nodules or as fine-grained concretionary masses, sometimes associated with vertical calcite veins and veinlets. Important pyrite accumulations closer to or on the surface are absent, because all the material is entirely oxidized to hematite, goethite and limonite.

3.3. Nonsulfide Minerals

The WMB nonsulfide deposits consist of a wide variety of minerals (Tables 4 and 5): barite, hemimorphite, willemite, smithsonite, zincite, hydrozincite, sauconite, cerussite, hydrocerussite, anglesite, wulfenite, pyromorphite, plattnerite, plumbonacrite, descloizite, vanadinite, hollandite, romanechite, cryptomelane, pyrolusite, hematite, limonite, and goethite. Such a varied mineral assemblage is not common in deposits connected to limestones or dolostones. Within the nonsulfide assemblage, willemite is the first mineral to be deposited, by direct replacement on sphalerite. Smithsonite partially replaces willemite and locally replaces hemimorphite, whereas hemimorphite replaces coarse-grained willemite crystals. Therefore, smithsonite and hemimorphite, in particular, signify the final generations of the nonsulfide mineral assemblage. The mineral assemblage at the WMB prospects varies from overbearing zincite, hemimorphite, smithsonite, and cerussite on/or near the surface, to a mixture of smithsonite, hemimorphite, willemite, and zinc clays in deeper zones (up to 50-70 m below the surface, or deeper). Massive willemite ore bodies are assumed to be available in the lowest part of the oxidized supergene deposits. Possible sources of SiO_2, required to form hemimorphite and willemite, are the underlying weathered igneous rocks and/or the silicified limestones and dolostones, as well as shale material. Significant amounts of cerussite were found in association with the zinc mineral assemblage, with Pb contents up to 8 wt.%.

Figure 4. (a) Mn oxides and zincite patches on barite. (b) Association of hemimorphite-willemite-zincite. (c) Dogtooth calcite crystals developed on hemimorphite.

The minerals occur in several structures, such as: shear zones, fractures, vugs, cavities, brecciated veins, and open spaces between the sidewalls of the barite veins, all connected mainly to the upper portion of the Cretaceous dolostone sequence. Sometimes they are connected to drusy, vuggy and brecciated structures, and are mostly associated with Fe-Mn oxides and/or hydroxides (Figure 4a).

(Nonsulfide deposits are characterized by variable zinc contents (Table 3), from low (1-8 wt.%), through medium (13-24 wt.%), to high grades 30 to 50 wt.%). Zn is contained mainly in hemimorphite, willemite, smithsonite, hydrozincite and zincite minerals. Pb contents are generally ≤5% (Table 3) but may reach up to 20 wt.% in neoformed galena, cerussite, hydrocerussite, wulfenite, anglesite, pyromorphite, plattnerite, and plumbonacrite. In addition, very high tenors of Pb are present in the silicified limestones, dolostones, and siliclastic beds and vein material. All the mineralised samples analysed are relatively Fe-poor varieties, with Fe contents lower than 3.4 wt.% (Table 3). Ag, Cd, and Ga are moderately high, and Mo may reach 1300 ppm. A very interesting issue is the content of Cr in these prospects, which is highly anomalous and may reach 6800 ppm.

Table 3. Chemical composition of nonsulfide-rich ore materials from WMB prospects

Element wt.%	MAM-2032	MAM-2033	MAM-2062
Zn	23.60	29.70	47.70
Pb	1.64	5.23	4.30
Fe	1.14	1.49	3.37
Mn	0.29	0.01	0.15
Mg	0.14	0.29	0.14
Ca	12.45	13.60	0.60

3.3.1. Hemimorphite

Hemimorphite (Figure 4b) has been recognized in almost all the WMB prospects, but more frequently in the southeastern part. Massive and/or fine-grained euhedral hemimorphite is more abundant than other nonsulfide zinc minerals (willemite, smithsonite, zincite, hydrozincite, etc.) in the oxidized zones (Alkohl, in particular; Table 4).

Hemimorphite is also reported from cavities in association with supergene brecciated material along with willemite, zincite and hydrozincite (Figure 4b). Hemimorphite may form transparent, lustrous and colourless radiating thin bladed crystalline clusters; massive reddish brown constituent of the matrix of supergene breccias; granular, white botryoidal and reniform aggregates resembling smithsonite; fibrous, finely needle-shaped and in many cases fan-shaped crystalline clusters; fine-grained thin bladed tabular transparent crystals freely growing in vugs, cavities, or at the contact with the barite veins.

Table 4. Mineralogical composition of hemimorphite-rich specimen from the Alkohl1 ore prospect

Mineral	Ideal Formula	Percentage
Hemimorphite	$Zn_4(Si_2O_7)(OH)_2 \cdot H_2O$	44.5
Smithsonite	$ZnCO_3$	21.3
Willemite	Zn_2SiO_4	11.7
Hydrozincite	$Zn_5(CO_3)_2(OH)_6$	11.6
Barite	$BaSO_4$	3.8
Zincite	$(Zn,Mn^{2+},Fe^{2+})O$	3.2
Calcite	$CaCO_3$	3.2
Maghemite	γ-Fe_2O_3	0.5
Quartz	SiO_2	0.2

3.3.2. Willemite

Willemite (Table 5; Figure 4b) has been detected in several samples from the WMB prospects, though it is less significant than hemimorphite, particularly at the upper levels of the nonsulfide occurrences. In vugs, willemite occurs as needle-like prismatic hexagonal crystals ranging in size from microscopic to several millimetres, usually forming aggregate bundles or radiating rosettes. Some drusy cavities are lined by radiating aggregates of colourless transparent, acicular willemite microcrystals.

Willemite becomes more important in the lower levels of the prospects. It occurs either as massive constituent, or as very fine-grained transparent, white, pale yellow, or pale brown to pale red euhedral crystals, or as reddish-brownish botryoidal concretions, generally smaller than 1 mm, but it may reach 1 mm in length, mainly in fissures, vugs, and open cavities, where the willemite crystals are affected by iron oxides coated on the surface. A predominance of massive and/or fine-grained crystalline willemite over hemimorphite and smithsonite is recorded in brownish to reddish colour veins. Relicts of radial white willemite crystal aggregates were also detected.

3.3.3. Smithsonite

Smithsonite has been identified in association with galena and cerussite, particularly at the external surfaces of the barite veins, or in fractures, vugs, cavities, and in association with massive nonsulfide Zn minerals. It is common in the upper parts of the mineralised WMB prospects, as botryoidal masses, in compact or granular forms and as sheaf-like aggregates. Macroscopically, smithsonite is white, brown, or grey, sometimes yellow tinted and rarely green or blue.

Table 5. Mineralogical composition of willemite-rich specimen from the Alkohl1 ore prospect

Mineral	Ideal Formula	Percentage
Willemite	Zn_2SiO_4	76.1
Hemimorphite	$Zn_4(Si_2O_7)(OH)_2{\cdot}H_2O$	14.5
Quartz	SiO_2	4.3
Cerussite	$PbCO_3$	2
Calcite	$CaCO_3$	1.1
Siderite	$FeCO_3$	0.9
Galena	PbS	0.8
Smithsonite	$ZnCO_3$	0.5
Total		100.20

3.3.4. Zincite and Hydrozincite

Zincite is found in association with hemimorphite and willemite (Figure 4b). It is usually massive, sometimes in granular form or in compact rounded masses. Its colour is orange red. Occasionally, it may be found as hexagonal fine-grained crystals developed on barite or on limestones. Hydrozincite is mostly found in association with nonsulfide Zn rich veinlets, developed as fine-grained crystals or massive granular aggregates of grey colour.

3.3.5. Calcite

Calcite crystals are very common, mostly colourless or tinted, variable in size, found as thick tabular, slender or needle-like, microcrystalline, or even as dogtooth spar developed as latest generation on descloizite (Figure 4c). Calcite occurs as fracture filling veins, as well as cavity filling material.

3.4. Vanadium Ores

Vanadium concentrations occur either in association with lead-zinc ores as vanadate minerals, or as anomalous traces in fissures of the Cretaceous dolostones. Descloizite and vanadinite mineralisations are oxidation products, representing the latest stage of formation, which took place when metavanadate-rich solutions were introduced into the deposit along near vertical fractures. Types of occurrence exposed on or near the surface are: structurally controlled veins and allied lineaments, as well as cavities; stockwork veinlets connected to calcite, also developed as massive descloizite specimens and patches on the sidewalls; disseminations found as crystal aggregates on the external contact of the barite veins as well as on the mechanically formed dolostones; and in association with siliclastic debris material.

3.4.1. Descloizite

The WMB prospects contain one of the most remarkable occurrences of descloizite known, and descloizite is probably the mineral resource for which it is most promising. Descloizite is associated with barite, quartz, calcite, Pb-V oxide, szomolnokite, supergene Zn-Pb carbonates, Zn silicates, and lesser sulphur (Table 6). Descloizite occurs in crystals with different shapes, including tabular, pyramidal, and prismatic. A youngest generation of tiny dogtooth spar crystals is often developed on the crests of previous descloizite crystals. The colours vary from metallic black, to dark green (Figure 5a), to olive-green, and to pale-brown and pale-red, and rarely golden.

Most descloizite specimens have metallic lustre, and the finest crystals are very highly lustrous. Crystal sizes range up to 3 mm in shallow zones, but they may exceed 1 cm in deeper areas. Some drusy aggregates (Figure 5b) may attain several kilograms in weight. Vanadium content of descloizite ore bodies reaches 12.30 wt.%, with anomalous Ag, Cd, Co, Cr, Ga, Mo, Ni, and Sb contents, whereas Cu is rare. High vanadium contents are reported in siliclastic materials (up to 0.7%), silicified dolostones (up 0.2%), and in Mn ores (0.1-0.2%), hollandite in particular. In addition, moderately anomalous V values (0.03-0.05%) were detected in limestone-rich bituminous material.

Table 6. Results of quantitative phase analysis (wt.%) of massive descloizite ore specimen from Alkohl1 prospect

Mineral	Ideal formula	Percentage
Descloizite	$Pb(Zn,Cu)(OH)VO_4$	74.8
Lead Vanadium Oxide	$Pb_{3.5}V_{4.51}O_{14.75}$	21.4
Szomolnokite	$FeSO_4 \cdot H_2O$	2.9
Sulphur	S	0.9
Total		100.0

Figure 5. (a) Dark green descloizite crystal aggregates developed on barite. (b) Greenish to brownish descloizite crystal aggregates.

3.4.2. Vanadinite

Vanadinite crystals are small six-sided prisms ranging in size from less than 0.1 mm up to 1 mm, and having bright brownish, orange, and yellow colours. Vanadinite occurs in vugs and fissures of the limestone formation and as impregnations on calcite-rich material. It is found in a relatively deep oxidation zone (approximately 100 m below the surface), in association with the alteration zone of lead-zinc minerals.

3.5. Mn and Fe Ores

The WMB area is characterized by Mn and Fe minerals related to the contact of the Cretaceous sandstones with the Cenozoic shallow marine carbonates. They form several discordant irregular veins striking NW-SE, near horizontal or, in some cases, steep dipping. Mineral phases mainly consist of hollandite, romanechite, cryptomelane, pyrolusite, hematite and limonite (Table 7).

Mn ore bodies occur as small massive veins, stockwork breccia-filling material, or cavity and porous filling with botryoidal of oolitic structure. Residual hollandite and pyrolusite fragments (up to 20 cm in diameter) are scattered and associated particularly to the debris carbonate sediments. In some cases, cryptomelane and pyrolusite occur as stockwork mineralisation cementing brecciated carbonate rocks. The manganese ores impregnatively penetrate the sandstones to a considerable depth. Further manganese anomalies were found in the metasomatically altered Cretaceous dolostones, along the active fault system connected to the Red Sea-Gulf of Aden rift.

Table 7. Results of quantitative phase analysis of manganese ores (wt.%)

Mineral	Ideal formula	MAM-432 (wt.%)	MAM-436 (wt.%)
Hollandite	$Ba(Mn^{4+},Mn^{2+})_8O_{16}$	100	27
Romanechite	$(BaH_2O)_2(Mn^{4+},Mn^{3+})_5O_{10}$	-	45
Quartz	SiO_2	-	18
Anhydrite	$CaSO_4$	-	9
Total		100	99

Results of chemical analyses of pyrolusite indicate high quality material with MnO >61% (Table 8). Ba and Sr may exceed 6% and 1.2%, respectively. Moderate anomalies of Co, V, Mo, U, W, Zn and Ag contents are reported from hollandite, romanechite and cryptomelane, indicating a genetic association with the hydrothermal mineralisation found elsewhere throughout the WMB platform. Possible sources of manganese are the Quaternary igneous mafic rocks, rich in Mn, where the released manganese primarily solves and migrates as hydro-carbonate, rarely as sulphate-complex forms.

Table 8. Chemical composition of representative samples of pyrolusite

Element wt.%	P1	P2
SiO_2	0.30	0.62
TiO_2	0.01	0.01
Al_2O_3	bdl	0.07
Fe_2O_3	bdl	0.04
MnO	61.61	61.38
MgO	0.66	1.00
Na_2O	0.41	0.44
K_2O	0.24	0.25
P_2O_5	0.05	0.06
(SO_3)	0.86	0.62
LOI	14.24	14.64

bdl = below detection limit.

Iron ores include several discordant but regular veins striking WNW-ESE, near horizontal or in some cases steep dipping, occurring in an intensively silicified material. The mineralogy consists of hematite, limonite and goethite (Figure 6). Sometimes the veins of iron ore are associated with barite, where barite may form 50% of the bulk. The iron ores occur as: massive veins attaining 2 m in width and more than 100 m in length; hematite-barite and hematite-barite-silicified dolostone stockwork breccia material; and cavity and porous filling types, characterized by botryoidal structure (Figure 6). Colour is dark grey, reddish brown, red or black. Limonite, found as pseudomorph after hematite, has yellow-brown and yellow colours, and usually forms botryoidal masses, and sometimes earthy. Manganese is often associated with hematite and limonite. Several analyses of hematite specimens yielded an average Fe content >55%, indicating a good quality iron ore. In some cases, anomalous V contents up to 0.3% are reported. Hematite occurs in different textures, either as patches on rosettes barite and quartz specimens, or as individual hematite-rich boulders.

Figure 6. Botryoidal goethite.

4. Discussion

The economic minerals of the WMB prospects occur almost entirely in the residual blanket, which overlies the late Proterozoic basement rocks. The mineralised zones usually occur in the upper portion of the sedimentary sequence (Saa'r Formation). They are commonly disseminated in a variety of structures such as solution breccias and pipes, shear zones, fractures and cavities.

Barite veins, which are mainly hosted in sedimentary bedded planes, are usually associated with near horizontal dolostone highs. Barite was formed by effect of hydrothermal and/or weathering activity, generally deposited in WNW-ESE trending faults, fractures, joints, as well as in planes of the bedded limestones and dolostones. Fossil records suggest the mineralisation occurred during Cretaceous time and continued up to the post-rift phase of the Red Sea–Gulf of Aden rift system (Beydoun, 1964).

Hypogene sulphide mineralisations consist of hydrothermally brecciated veins and cavity filling material, settled in the deformed limestone and dolostone strata of Jurassic and Cretaceous ages, possibly connected to a stratabound structure. The mineralised veins consist of sphalerite, galena and pyrite. Sphalerite was found only in the deeper parts of the silicified veins, indicating higher temperature of formation, whereas galena occurs in the upper parts of the barite and silicified veins, indicating lower temperatures. Important pyrite accumulations closer to or on the surface are absent, because all the material is entirely oxidized to hematite, goethite and limonite.

Nonsulfide Zn–Pb ores consist of a supergene mineral assemblage, with mostly hemimorphite, willemite, smithsonite, hydrozincite, and zincite. Evidence for the formation of Zn-Pb nonsulfide deposits indicates that oxidising conditions had reached considerable depths below the surface, probably up to 150 m deep. In addition, formation of willemite can be ascribed to hydrothermal processes connected to the Quaternary volcanic activity. The preponderance of hemimorphite in the upper levels of the WMB nonsulfide occurrences indicates the persistent replacement of willemite, and this is in accordance with the typical mineral assemblage of hemimorphite and/or smithsonite found in most European supergene nonsulfide ores (e.g., Boni and Large, 2003). Willemite becomes more significant in the lower level deposits, due to the achievement of temperatures exceeding 100°C and oxidizing fluids. It is intensively replacing sphalerite, since willemite is more stable at higher temperature as well as at lower water activities. In addition, willemite overcomes smithsonite at higher activity of silica, for which a likely origin is hydrothermal alteration of igneous rocks.

Vanadium ores, consisting of descloizite and vanadinite, are either breccia type occurrences, or metal-enriched silicified dolostones as well as siliclastic layers near the surface, characteristic of semiarid regions. Based on depositional features and mineralogy, descloizite post-date the barite mineralisation suggesting that it was formed in a relatively late phase. The occurrence of descloizite in close proximity to other nonsulfide mineral assemblages and/or in partial association with them, as well as its decreasing amounts with decreasing depth, suggests a supergene genetic model for the formation of descloizite. Geochemically, vanadium has a lithophile character, but under the action of oxygen-rich, acidic waters in semi-arid and arid climates, it dissolves as oxyanion vanadate (VO_3^-) or cation vanadyl (VO_2^+). In such oxidation zones vanadium concentration follows mechanisms similar to those forming "laterite infiltration" of iron. However, the perfect crystalline form of several descloizite generations and their common connection with barite (Figure 5a), calcite and silicified dolostones, may support an origin by hydrothermal activity, in which descloizite might be associated with fluids derived from the Cenozoic basalts. Indeed, these lavas are rich in vanadium (200-300 ppm V; Mattash et al., 2014 and references therein) and can be considered an important source of this element.

The polymetallic deposit at the WMB was formed and concentrated by effect of hydrothermal processes connected to the Cenozoic-Quaternary volcanic eruptions, and/or by the action of slightly acidic groundwater which dissolved the soluble carbonate during a long period of weathering and erosion. It is also possible that the WMB mineral prospects were accumulated during earlier cycles of weathering closely related to peneplains, which were later uplifted and dissected owing to the opening of the Red Sea–Gulf of Aden rift system.

CONCLUSION

The common mineral association in the WMB mineralised prospects is barite, galena, sphalerite, willemite, zincite, hydrozincite, smithsonite, galena, descloizite, and less commonly vanadinite. Paragenetically, it appears that willemite formed first, with zincite, hydrozincite, smithsonite as well as hemimorphite following in determined order. Calcite is a widespread gangue mineral. Mn and Fe oxides and hydroxides are common and include hollandite, romanechite, cryptomelane and pyrolusite, as well as hematite, limonite, and goethite, respectively. The ore bodies are extensively oxidized and enriched in nonsulfide Zn and Pb minerals and breccias. In addition, chalcedony and jasper occur as latest generation, together with fine-grained transparent hemimorphite crystals. Dolomite, aragonite, siderite, goethite, limonite, gypsum, and clay minerals are also reported. Euhedral coarse, medium and fine-grained quartz crystals occur as cavity or drusy filling.

The mineralisation style at WMB depends on both the facies and the diagenetic characteristics of host rocks. Mineralisation also depends on tectonic setting (structurally controlled) and the occurrence of sedimentary structures, such as joints, cavities, and vugs, as well as on tectonic breccias, along the mineralised fractures in particular.

The ore minerals were formed and concentrated by hydrothermal processes caused by the circulation of fluids produced by the Cenozoic-Quaternary volcanic activity, and/or by the action of slightly acidic groundwater which dissolved soluble carbonates during a long period of weathering and erosion. Sources of the mineralizing fluids are the Wadi Masilah main drainage system (370 km long) and its parallel lineaments, as well as the volcanic activity that occurred along the southeastern part of WMB. Tectonic uplift and intensive fracturing system, along with favourable climatic conditions, gave rise to the development of supergene deposits, including Fe, Mn, Pb, V, and Zn.

Based on our geological and geochemical investigation, the WMB is highly prospective for 'Small scale' barite, as well as important stratabound lead, zinc, iron, and manganese deposits. In addition, the WMB contains remarkable occurrences of descloizite, and it is very promising for this mineral resource.

Further detailed geological, geochemical, and geophysical surveys, and drilling works, will be necessary in order to enhance the ultimate economic potential of such significant mining district.

Appendix: Analytical Methods

All the samples collected were processed and prepared at the Geological Survey and Minerals Resources of Hadramawt, Mukalla (Yemen). The identification of mineralogical species was achieved by X-Ray Diffraction analysis at the University of British Columbia, Canada. Chemical analysis of powdered samples was performed in part at ALS Chemex Laboratory, Vancouver (Canada), and in part at the Geochemical Laboratory of the University of Hannover (Germany). Analytical methods used are X-Ray Fluorescence for major oxides, Inductively Coupled Plasma Spectrometry for trace elements.

Acknowledgments

The author is grateful to Federico Pezzotta for constructive comments on an earlier version of the chapter.

This work was financially supported by MEMC, Naine Minerals and Energy and Mining (AJHAM) Companies, by the Geological Survey of Yemen, and by the National Research Council, Institute of Geosciences and Earth Resources (CNR-IGG), Section of Florence (Italy).

REFERENCES

As-Saruri M., Sorkhabi R. (2016) Petroleum basins of Yemen. *Geo Ex Pro* 13(2), 38-42.

Beydoun Z. R., (1964) The stratigraphy and structure of the eastern Aden Protectorate. Overseas Geology and Mineral Resources. *Bull. Suppl. Overseas Geol. Min. Resources* 5, 107 pp., H.M.S.O., London.

Beydoun Z. R. and Greenwood, J. E. G. W. (1968) Aden Protectorate and Dhufar. In: Dubertret L. (ed.), *Lexique Stratigraphique International Vol. III. Asie*, 1062 pp., CNRS, Paris.

Beydoun Z. R., Bamahmoud M. O., Nani A. S. O. (1993) The Qishn Formation, Yemen: lithofacies and hydrocarbon habitat. *Mar. Petrol. Geol.* 10, 364-372.

Beydoun Z. R., As-Saruri M., Baraba R. S. (1996) Sedimentary basins of the Republic of Yemen: their structural evolution and geological characteristics. *Rev. Inst. Franc. Petrol.* 51, 763-775.

Beydoun Z. R., As-Saruri M. L., El-Nakhal H., Al-Ganad I. N., Baraba R. S., Nani A. O., Al-Aawah M. H. (1998) Republic of Yemen. *International Lexicon of Stratigraphy. Vol. III, Asia, Fascicule* 10b2, IUGS Publ. 34, 245p.

Boni M., Large D. (2003) Nonsulfide Zinc Mineralisation in Europe: An Overview. *Econ. Geol.* 98, 715-729.

Bott W. F., Smith B. A., Oakes G., Sikander A. H., Ibraham A. I. (1992) The tectonic framework and regional hydrocarbon prospectivity of the Gulf of Aden. *J. Petrol. Geol.* 15, 211-243.

Ellis A. C., Kerr H. M., Cornwell C. P., Williams D. O. (1996) A tectonostratigraphic framework for Yemen and its implications for hydrocarbon potential. *Petrol. Geosci.* 2, 29-42.

Falagàn C., Grail B. M., Johnson D. B. (2017) New approaches for extracting and recovering metals from mine tailings. *Mineral. Engineer.* 106, 71-78.

Haitham F. M. S., Nani A. S. O. (1990) The Gulf of Aden Rift: hydrocarbon potential of the Arabian sector. *J. Petrol. Geol.* 13, 211-220.

Hughes G. W., Clarke M. W. (1988) Stratigraphy and rock unit nomenclature in the oil-producing area of interior Oman. *J. Petrol. Geol.* 11, 5-60.

Hughes G. W., Beydoun Z. R. (1992) The Red Sea-Gulf of Aden: biostratigraphy, lithostratigraphy and paleoenvironments. *J. Petrol. Geol.* 15, p. 135-156.

Mattash M. A. (1994) *Study of Cenozoic Volcanic Rocks and Their Associated Intrusions*. Ph.D. Thesis, University of Budapest, Hungarian Academy of Sciences. 112 p. (Unpublished Thesis).

Mattash M. A., Al-Ameri A. A, and Shaqra A. A. (2007a) Geology and geochemistry of barite prospect in Yemen. *Sixth International Symposium on Eastern Mediterranean Geology incorporating the Ninth International Conference of Jordanian Geologists Association*. Extended Abstract, p. 249-250.

Mattash M. A., Shaqra A. A., and Al-Ameri A. A. (2007b) Geological survey and feasibility study of pyrolusite deposit in Yemen. *Sixth International Symposium on Eastern Mediterranean Geology incorporating the Ninth International Conference of Jordanian Geologists Association.* Extended Abstract, p. 251-252.

Mattash M. A. (2008) Geology and geochemistry of the sedimentary-hosted lead-zinc-vanadium-barite prospects in Yemen. *International Geological Congress*, Oslo-Norway. Abstract v. 1p.

Mattash M. A., Al-Ameri A. A, and Shaqra A. A (2009) Geology and geochemistry of the carbonate hosted polymetallic-barite prospects in Wadi Al-Masila, Al-Mahrah province, Yemen. *Second Syrian Geological Conference*, Damascus, Syria. Abstract v. 1p.

Mattash M. A., Pinarelli L., Vaselli O., Minissale A., Al-Kadasi M., Shawki M. N., Tassi F. (2013): Continental Flood Basalts and Rifting: Geochemistry of Cenozoic Yemen Volcanic Province. *Int. J. Geosci.* 4(10), 1459-1466.

Mattash M. A., Pinarelli L., Vaselli O., Minissale A., Jaimes-Viera M.C., Al-Kadasi M., Shawki M. N., Tassi F. (2014) Geochemical evolution of southern Red Sea and Yemen flood volcanism: evidence for mantle heterogeneity. *Arab. J. Geosci.* 7(11), 4831-4850.

Paul S. K. (1990) People's Democratic Republic of Yemen: a future oil province. In: Brooks, J. (ed.) Classic Petroleum Provinces. *Geol. Soc. London Spec Pub*. 50, 329-339.

Powers R. W., Ramirez L. F., Redmond C. D., Elberg E. L. (1966) Geology of the Arabian Peninsula. Sedimentary Geology of Saudi Arabia. *USGS Professional Paper* 560-D, pp. 147.

Redfern P., Jones J. A. (1995) The interior rifts of the Yemen - analysis of basin structure and stratigraphy in a regional plate tectonic context. *Basin Res.* 7, 337-356.

Schlumberger (1992) Looking for Yemen's hidden treasure. *Schlumberger Middle East Well Eval. Rev.* 12, 12-29.

Reviewed by: Federico Pezzotta, Curator of the Mineralogical Section of the Natural History Museum, Milano, Italy

In: Red Sea
Editor: Delbert Robbins
ISBN: 978-1-53613-200-7

Chapter 3

Current Status of the Sudanese Red Sea Fisheries Resource Management

Omyma Karrar, PhD
Arab Open University, Kuwait City, Kuwait

Abstract

The Sudanese Red Sea is considered to be unique in many aspects. It is still pristine with rich and diversified ecosystems and high biodiversity in addition to other potential resources of high economic value. Among these resources is a rich fisheries reserve dominated by different assemblages of an ecologically and commercially important stock of invertebrates, demersal finfish and pelagic fish. The national fisheries standing stock, excluding the shared stock, remains largely under-exploited with the exception of Mother of Pearl (*Pinctada margeritifera*) and Trochus (*Tectus dentatus*) which are efficiently utilized. The fisheries resource if fully exploited will greatly contribute to the food security of the region and provide sources of income to the local communities. Official statements and reviews including international and regional reports by the Food and Agriculture Organization of the United Nations ((FAO), 1989), United Kingdom Overseas Development Administration

((ODA), 1996), Regional Organization for the Conservation of the Environment of the Red Sea and Gulf of Aden ((PERSEGA), 1998) and Sudan National Report (2013) have identified a number of challenges that the Sudanese Marine Fisheries Sector is facing in order to increase the catch yield.

Keywords: Sudanese Red Sea, Marine fisheries sector, standing stock, food security, catch yield

About this Chapter

The first section of this chapter will provide general information about the study area and outline the background of the issue. Next, the findings and key issues will be discussed including the constraints and challenges of the Sudanese Red Sea fisheries resource management. Finally, future directions including actions by local policy makers to address existing challenges both in terms of activities for improving the yield and measures to ensure sustainable use of the resource are provided.

Introduction

Background Information

The Sudan, officially known as The Republic of Sudan was the largest country in the continent of Africa until the separation of South Sudan in 2011. It is now the third largest country in Africa and 16th largest in the world, with an area of 1,861,500 km^2. Sudan lies between latitudes 8° and 23°N, within the North Eastern Part of Africa and is bordered by the Red Sea to the east, and seven countries. The bordering countries and their relative location are as follows: Egypt to the north, Eritrea and Ethiopia to the east, South Sudan to the south, the Central African Republic to the southwest, Chad to the west and Libya to the northwest. The country's terrain is generally made up of flat plains, interrupted by several mountain ranges, including the Marrah Mountains (3,042 m or 9,980 ft) in the western region, which form the highest point in Sudan, and the Red Sea Hills in the eastern region (Sudan Country Profile, 2017). The River Nile is considered to be the most significant feature in the country's topography, forming a prominent incision in the terrain and dividing the country into eastern and western halves. The River Nile, which flows northwards through Egypt and ultimately to the Mediterranean Sea, consists of the Blue, and White Nile rivers, which meet in Khartoum, the Capital of Sudan.

Sudan's population according to the most recent census reports, is 40,533,550, with an annual growth rate of 2.38% as reported by the United Nations Population Division for the year 2017 (United Nations, 2017). The country is rich in both living and non- living resources. The natural resources include reserves of crude oil, precious metals, significant animal wealth and massive areas of fertile arable land. Agriculture (including fisheries) being the backbone of the country's economy. Agriculture contributes 42 percent to the annual GDP, however, the contribution of the fisheries sector is relatively limited. Sudan's exports are minute and were estimated to be 0.2 million dollars in 2012. The country also relies on imports of fish and fishery products amounting to 5.2 million USD dollars.

Sudan's fisheries sector is known to have a rich resource base acquired from both marine and fresh water reserves. The marine resource includes offshore and inshore waters. The fresh water resource is mainly based on the River Nile and its tributaries, wetlands, lakes and man-made reservoirs. The capture fisheries' production was estimated to be about 34, 000 tons in 2012, 29,000 tons from fresh inland water catches and 5,000 from marine catches. Capture fisheries' activities are centered around the River Nile and its tributaries, seasonal flood plains and four major reservoirs, as well as the territorial waters of Sudan on the Red Sea (Country Brief, 2014).

It is important to note the heavy involvement of international Non-Governmental Organizations (NGOs) and several other agents including UN (United Nations) agents in the Sudan as result of the many years of civil war, existing internal conflicts in addition to a large influx of refugees from neighboring countries. It attempts to help alleviate the difficulties faced by displaced people and the country in general. Therefore they have a strong presence throughout the country. The NGOs and UN agents operating in Sudan include the World Food Program (WFP), the Food and Agriculture Organization of the United Nation (FAO), the United Nations Development Program (UNDP), the United Nations Industrial Development Organizations (UNIDO), the United Nations Children Fund (UNICEF), the United Nations High Commission for Refugees (UNHCR), the United Nations Mine Service (UNMAS), the United Nations Office for the Coordination of Humanitarian Affairs (OCHA), World Bank, and the International Organization for Migration (IOM). Although these organizations are primarily providing humanitarian aid, they contribute significantly to the development of the country through joint developmental projects, funds, technical support, and training, in addition to consultancies and expertise. Examples of the influence and involvement of these organizations in relation to Sudan's agriculture and fisheries resource management will be discussed in this chapter.

Study Area

Red Sea State: Sudan adopts a federal system and consists of 18 States, with the Red Sea State (Al-Baher Al-Ahmer) being one of them. It has an area of 212,800 km^2 and an estimated population of 1,396,000 (Sudan Country Brief, 2017). The Red Sea State is divided into nine administrative areas: Port Sudan- the capital of the state and the main port, Suakin- the second sea port, Gebiet Haya, Sinkat, Gunob Awlieb, Derodieb, Tokar and Ageeg. As the name suggests, the Red Sea borders the state to the east. Inland, mountains run from north to south, interrupted by arid plains. To the northwest of the state is the Nubian Desert. The Siyal Islands are located in the disputed area between Egypt and Sudan in the northeast.

The population of the Red Sea State consists of rural nomads, rural settlers and urban people. The original inhabitants of the state are the Beja tribes, who make up over 65% of the current population, In Sudan, the Beja, who are of Hematic origins, have been considered as indigenous and the largest ethnic tribe in the eastern state. Within the Beja tribes, the Amarar and Besharin sub-groups occupy the northern part of the Red Sea State and are both nomadic pastoralists and fishermen. The Hadandaw and Beni Amer sub-groups in the south Kassala state are agro-pastoralists and fishermen (Saeed, 2004).

The Sudanese Red Sea coast lies within the Red Sea State and extends for 750 km including embayment (Schoreder et al., 1980). The Exclusive Economic Zone is 91,600km^2 (square Kilometers) of which 22,300km^2 are fringing reefs. The major features include a lining of fringing reefs, coastal lagoons, barrier reefs and atolls. Several of these lagoons are fringed by mangroves (Khalil & Krupp, 1994), which along with coral reefs support many fisheries. The productivity of the Red Sea is relatively poor as it is a semi-enclosed water body, however biodiversity and species endemism are quite high. It is estimated that between 10% and 30% of the Red Sea fish are not found anywhere else (Randall, 1992). The potential of the Red Sea resources is fairly high and levels of exploitation are generally low. This fact was noted by member countries bordering the Red Sea, and with a sense of regional responsibility, those countries formed a regional coalition to assess the fisheries resources status and address relevant challenges. The outcome has since been a Strategic Action Plan which was formulated as an immediate need for the Red Sea Countries with vested interest in promoting regional development (PERSGA, 1998).

Capture fresh water fisheries were found to be the main contributors to the total national production, whereas the Marine fisheries contribution was determined to be relatively small (Farah, 2000). Overall, it is frequently reported that the Marine fisheries resource utilization is not enough to satisfy the national demand, and the demand had to be partially met by imports, often resulting in lower average consumption per capita rates as the prices of fish products are generally higher than meat. The country is also dependent on imports of fish and fishery products (USD 5.2 million in 2012) to satisfy the limited per capita fish consumption (about 1.1 kg in 2012). Exports on the other-hand are insignificant and were valued at 2 million USD in 2012 (Country Brief, 2014).

These exports are mainly attributed to marine fisheries products such as finfish, Crustacean, Mollusca, and sea cucumber in addition to limited amounts of processed food (dry and wet salted fish). Some of the Marine Fisheries products are exported regionally to some Arab countries and neighboring African countries. Some products are exported to Europe and Asia.

Figure 1. Map of Sudan States including the Study Area (Al Bahr al Ahmar State – The Red Sea State).

METHODS EMPLOYED

This chapter aims to critically review the current state of the resource management activities of fishers in the Sudanese Red Sea by providing an account of the fisheries resources base, reviewing the current status of fisheries resource management, and assessing the role of local policies in both addressing key challenges, and ensuring sustainable utilization of the resources.

The assessment and evaluation entailed intensive literature review covering the relevant aspects of the subject matter, review of technical reports, national and international official records, in addition to personnel communication and the author's expert knowledge from previous experience and research in the Sudanese Red Sea.

FINDINGS

The following summaries and tables reflect the status of the Marine Fisheries Resource over a period of time (two decades). The main aim is to illustrate the progress of the sector and to evaluate the current status and management of the resource.

Resource Base and Fishing Activities

The Sudan Marine fishery is divided into three sectors i.e., commercial, artisanal and subsistence. The most dominant activity is the artisanal fishery. In general terms, the marine fishery may be classified into offshore and inshore (commercial and coastal artisanal) fisheries. The inshore waters, which lie from five nautical miles off the coast up to the edge of the continental shelf area, are exceptionally rich in resources and have a very active fishery.

Table 1. The most commonly fished species of The Sudanese Red Sea & the relevant basic information

Scientific name	Basic Information
Cartilaginus Fishes	
Galeocerda Cuvier (Tiger Shark)	• Twenty three species of shark representing seven families are recorded in the Sudanese Red Sea. • Sharks are not efficiently utilized and signs of over exploitation are reported. • Shark production was 2340 tons in 2009. The annual production decreased to 26 1 tons according to the most recent estimates of 2013.Shark fishing is currently restricted by a Ministry decree. A conservation & management plan for sharks' fishery is being carried out by the State Authorities.
Ginglymostoma ferrugieeum (Rusty shark)	
Alopias vulpinus (Thresher shark)	
Mustelus cuvier (Smooth dogfish)	
Scoliodon palasorrah (Dog shark)	
Bony Fishes	
Lutajanus bohar (Bohar)	• A total of 236 species of bony fish have been reported in the Sudanese Red Sea. • The species highlighted in this table constitute 60-70% of the finfish catch due to abundance and customer preference. • The Potential yield is estimated at 10 000 tons/year, while the reported yield amounts to 5000 tons/year. • *Pelctropomus macultus* known locally as Najiil is a key market species for local consumption and regional export. It is more heavily fished than others.
Lutjanus gibbis	
Lethrinus spp (Shaoor)	
Caranx spp(Bayad)	
Pelctropomus maculatus(Najiil)	
Scomberomorus commerson	
Mugil spp (Gushar)	
Epinephallus aerolatus	
Apprion spp (Farisi)	
Crustaceans (Shrimps)	
Peneaus semisulcatus	• A total of eight species of shrimp have been recorded in the Sudanese waters. The species highlighted in the table form the bulk of the harvest due to their high commercial value. • They form a high percentage of the Marine resources exported abroad because of the high demand for shrimp. • The crustacean resources have not been properly quantified. Estimates report the potential yield to be 20,000 tons. • The current ceiling for maximum catch is 1500 tons annually. • A limited range shrimp culture is taking place in the southern region of the Sudanese Red Sea. • Shrimp and prawns are also sold locally as a highly-valued delicacy food, particularly in Khartoum, the capital of Sudan.
Peneaus latisulcatus	
Metapeneaus monocerus	

<table>
<tr><th>Scientific name</th><th>Basic Information</th></tr>
<tr><td colspan="2">Mollusca</td></tr>
<tr><td>Pinctada margaritifera
(Mother of pearl)</td><td rowspan="4">• Diving in search of wild mollusca represents an ancient occupation for the majority of the coast population.
• Shells of the mother-of-pearl oyster Pinctada margaritifera and the gastropod Trochus dentatus are exported to some European countries.
• They contribute to 90% of the wild mollusca collection.
• The estimated annual landing for trochus is 500 tons and 30 tons of Pinctada.
• Strombus and Lambia shells are harvested and sold locally as a source of calcium, food and as souvenirs.
• Estimated annual landing for Strombus is 600 pieces. Lambia constitutes 500 kg of the landing.
• Strombus is also used as a base of local perfume manufacture (Dufra). This resource is not well managed and signs of depletion are reported.</td></tr>
<tr><td>Trochus dentatus (Kokian)</td></tr>
<tr><td>Strombus Gigas
(The queen conch)</td></tr>
<tr><td>Lambia spp</td></tr>
</table>

The artisanal fishery uses traditional fishing methods and gears. These include hooks and lines for capturing reef fish, where most of the fishing activities are concentrated. The common fishing crafts used are small boats of lengths ranging between seven and eight meters (7-8m) and are mainly boats constructed of wood or fiberglass. Several larger launches also operate in the southern region of the Sudanese Red Sea.

Their lengths range between 9 meters and 13 meters, and usually fish for Mollusca (Trochaus and Strombus; FAO, 1984; UNIDO Inception Report, 2009). Subsistence fishing activities on the other hand are very primitive and localized, and are generally conducted using more basic fishing methods such as spears, traps, cast-nets, and hook-and-line. The commercial marine fishery uses more sophisticated gear and crafts, i.e., better equipped motorized boats, trawlers, purse seiners. It requires sufficient resources and is therefore carried out by few fisher cooperative societies and independent private sector groups (Abeywickrama, 2009).

The commercial marine fishery focuses on both pelagic (occupying the pelagic zone up to 600 feet below the continental shelf) and demersal finfish (inhabiting the sea bed and zones near the continental slopes) such as groupers and snappers, e.g., threadfin bream, lizard fish and goat fish, as well as on shrimp resources. However, it is largely seasonal in nature (May to December due to weather conditions). On the other hand, the artisanal finfish fishery targets Pelagic fish such as Grey Mullet, *Epinephelus Areolatus*, *Lutjanus Bohar*, *Scomberomorus Commerson*, Sharks and Molluscs (Table 1). The Sea Cucumber (Echinoideo) fishery and the Sardine fishery constitute a small portion of a total marine catch of 5 700 tons reported in 2009. Recent records reflect a moderate increase in the overall catch (Abdulla, 2013.)

Shrimp constitutes the bulk of the harvest of crustaceans which are caught, mainly *Peneaus Semisulcatus, Peneaus Latisulcatus*, and *Metapeneaus Monocerus* species. The fishing grounds are concentrated in the southern coast of the Sudanese Red Sea, more specifically, Toker Delta, Gulf of Agieg, Mersa Mogadam, Khor Nawarat and Suakin.

Pinctada Margaritifera (mother of pearl), *Trochus Dentatus*, *Strombus spp.* and *Lambia spp.* are the principal Molluscs the fishery harvested. They have a high commercial value due to a large percentage of export to Europe and other international markets. *Pinctada Margaritifera* (the mother of pearl) is abundant in the northern zone of the Sudanese Red Sea mainly Dungonab Bay and Um EL-Sheikh Island (Figure 2). Diving in search of mother of pearl is one of the oldest activities of the local communities of the Red Sea region. It is also the principal species in Sudan Mari culture since 1905. The initiation of the oyster culture industry was based on 'Cyral Crossland's efforts as one of the pioneer scientists working in the Red Sea area (Crossland, 1957). A number of family farms were established after the piloted oyster farms had proven successful. The main objective was to increase oyster production for the international market. However, the oyster culture industry was affected by two episodes of mass mortalities in the years 1996 and 1973. In recent years, great efforts have been directed towards reviving the oyster culture by the Fisheries Administration. To this effect, a joint program was initiated by the Ministry of Animal Resources and Fisheries of the Red Sea State and support from two international organizations, namely ACCORD (African Centre for the Constructive Resolution of Disputes) and OXFAM (Oxford Committee for Famine Relief). Alternative culture technologies are employed instead of the previously adopted conventional practices. The program aims to revive and encourage the expansion of oyster family farms along the coast that will contribute to improving the economic status (UNIDO Inception Report 2009; Sheikeldin, 2014).

Figure 2. Important Localities and Landing Sites of Fishing Activities.

The Strombus is highly valued and harvested locally. It is used as a source of food as well as a base for local perfume manufacture, therefore it is heavily exploited (Schroeder, 1981 & PERSEGA, 1998). In addition, recent country reports (2014) estimate that the average sale of Dufra ranges from 13-15 pounds/day, where each pound produced contains the opercula of 1000-2000 depending on the size of the gastropod animals (Altayeb, 2001).

According to national records, total marine fish production was 369 metric tons excluding sharks which constituted 11 metric tons, i.e., 2.9% of the total fish production. Saeed (2004) reports that the marine fin fish production constitutes 7% of the total estimated fish potential of the Marine sector which is estimated to be 35 000 tons/year. However in 2011 the total yield was reported to be 5700 metric tons (Abdullah, 2013).

Despite the rich resource base and the great economic potential, the marine fisheries sector is still not fully developed and the annual production is far below the Maximum Sustainable Yield (MSY). This status is attributed to challenges and constraints that the fishery sector is facing.

Challenges and Constraints

The table below highlights the challenges and constraints and the potential impacts.

Table 2. Major Constraints & Challenges Affecting the Marine Fisheries Yield

Constraints	Impacts
Poor records and data collection methods	Affect the overall resource planning and management
Inadequate infrastructure (roads, transport, cold storage, equipped landing sites, etc.)	Lead to many post- harvest losses of the fishery yield
Financial constraints	Limited support for fishing activities leading to underutilization of the resource
Dated policies & regulations	Restrict proper implementation, monitoring and surveillance. Over exploitation of specific fisheries, illegal fishing. Non Compliance
Limited institutional support	Weak management practices

Constraints	Impacts
Weak organization of fishing communities	Low income and poor socio economic status. Weak representation and support for the fishers
Lack of clear marketing strategies	Yield losses, low returns, monopoly by few groups of traders
Inadequate technological resources and technical expertise	Impacts the overall fishing activities. Improper data collection and analysis, monitoring and surveillance, handling and processing of the fisheries yield
Limited research capacity	Direct effects on eco-management and conservation of the resource base
Natural & environmental challenges	Limits resource utilization and management, leads to environmental degradation

Poor Documentation and Data Collection Methods

Proper planning and management requires current, consistent and accurate statistics and data of the resource base, and proper documentation systems. As mentioned earlier in the introduction, fish stock assessment studies are mostly fragmented and rather limited. There is a general lack of detailed stock assessment and lack of adequate time series that greatly impact the overall resource management practices. Data collection methods are far from comprehensive. They are limited and mostly restricted to landing estimates on landings and efforts. In addition, very little biological information is available such as distribution and interdependence of many of the major commercial species. Lack of comprehensive data and detailed information is considered to be one of the major challenges that requires immediate attention.

Inadequate Infrastructure

The Sudanese Red Sea State is one of least developed regions with the exception of Port-Sudan, the capital of the state and Sudan's main national port. Large areas of the coast lack basic physical infrastructure such as, paved roads, proper transportation, and communication systems. In addition, many fish landing sites are not provided with basic amenities.

There are a total of 36 boat landing sites along the coast, only a few of which are equipped with proper facilities and support. As a result, post-harvest activities are very much affected leading to great losses of the fishery's yield.

Financial Constraints

Limited governmental financing is the main constraint that drastically affects the marine fisheries sector throughout the different stages of the resource utilization cycle. Sources of income are limited to the allocated national budgetary allowance for the fisheries administration. Other sources of income include the returns from craft licensing and registration and work permits from fishers and the respective working entities within the sector.

Limited Institutional Support

At a federal level, the responsibility for fisheries management rests with the designated fisheries authority, i.e., the Ministry of Animal Production and Fisheries Resources. At an institutional level, the Fisheries Administration reports to the undersecretary of the Ministry and conducts its duties as per the Central Ministry mandate, strategy and action plans. The Fisheries Administration also coordinates its activities with the relevant institutions, including the Marine Research Center (MFRC), Ministry of Environment and Tourism, and the Sudan Marine Conservation Committee (SMCC). The main objective of this administrative framework is to increase the overall fish yield and improve its marketing and distribution, promote investment and develop fishing communities, as well as coordinating efforts for integrated coastal and marine management at national, regional, and international levels (Sanders & Morgan1989) (Sudan Country brief, 2014). The fishery administration of the Red Sea follows a similar organizational and administrative framework but at a local state level. However, the organizational framework clearly sets the administrative and functional reporting boundaries. Like other governmental units, the fisheries administration has limited financial resources. The contribution from the private sector is also limited and is mainly based on registration fees, licenses and permits of the fishery sector. The State Fisheries Administration supports the activities of the two sub–stations (Fisheries research and Aquaculture research). Although the Fisheries Administration has action plans in place, limited finance and capacity building restrict implementation activities.

Weak Organization of Fishing Communities

The fishers' body constitutes a total of 2,300 registered persons, working in the coastal marine fisheries. The majority work on a fulltime basis. Some of those fishers have undertaken fishing as an alternative mean of livelihood instead of their original farming or herding activities due to harsh environmental conditions, i.e., draught spells, desertification and water shortages that greatly affected their subsistence activities. Few others are engaged in other part time activities to diversify their limited sources of income. The living conditions are generally poor and the income generated from fishing hardly satisfies their basic needs. However, more than anywhere else, the Sudanese fishers' livelihood depends largely on fishing activities as an income generation source and a major source for their daily food consumption. The fishers of the Red Sea State belong to various cooperatives associations and trade unions. The Government has established community based fishing associations and cooperatives to organize fishing activities and to provide the fishing communities with the required support. The introduction of the cooperatives dates back to the 1970s when a very limited number of cooperatives existed, and operated mostly in Port Sudan. Statistics from reports released in 2009 (Abeywickrama, 2009) concluded that the total number of cooperatives has increased to 30, expanding their support to the main fishing communities which are spread out in several areas along the northern and central stretches of the coast including Port Sudan, Suakin (the second port of the Sudanese Red Sea), and Ageig in the southernmost part of the coast. To the north of Port Sudan, the fishing communities are mostly concentrated in Dungonab, Mohamed Goul and Arkiyai. It is reported that the contribution of the cooperatives in supporting the fishers and developing the economy is below expectations. Most of the cooperatives have failed to address the fishers' pressing needs. The UNIDO inception report (2009) identified a number of reasons in this regard; mainly, the weak organizational structures of the cooperatives, the lack of management skills, the absence of detailed records, poor documentation and archiving systems, and limited trust and information sharing among the associations' working groups. However, it is important to note that the recent restructuring of the cooperatives and establishment of business units has been a great step forward in the empowerment of the fishing communities.

Lack of Clear Marketing Strategies

Successful business requires having a clear and transparent marketing strategy in place. Such marketing strategies should be based on comprehensive market assessments that address all respective aspects of the production cycle and marketing chain. A bilateral project titled "Recovery of Coastal Livelihoods in the Red Sea of Sudan through the Modernization of Artisanal Fisheries and the Creation of New Market Opportunities" was initiated in 2009 with the aim of addressing such marketing issues. The multifaceted project addresses, among other things, marketing strategies and policies. The main objective is to structure the fishing industry to alleviate fishers' poverty, given the fact that they are completely dependent on wholesale fish traders who provide them with the necessary working capital. It also aims to improve the economic status of the fisheries sector communities as well as to provide better fish marketing services to consumers while ensuring sustainable development of the fisheries resources. Consequently, micro-business support services in the larger fishing sites have been established. Two national entities mainly

'Zakat Chambers' and Agricultural Bank are providing loans to fishers to reduce their dependency on fish traders. Support includes fishing gear and crafts.

Inadequate Technological Resources and Technical Expertise

Gaining access to advanced technologies and developing technical expertise is an issue of vital importance and requires great improvement and development in the context of fisheries resource management in the Red Sea state of Sudan. The Marine fisheries sector increasingly rely on technological infrastructure in almost all aspects of utilization and management of the resource, starting from digital data and informatics through various computer programmed models for the purpose of stock assessment, modelling and management approaches. Fishing craft and fish industry monitoring and surveillance require specific technological infrastructure and technical expertise which are not always available, and the available ones are often obsolete.

Dated Policies and Regulations

The first national legislation to govern and regulate the marine fisheries was developed in 1937 as the "Marine Fisheries Ordinance." The ordinance is composed of eight articles that mainly address fishing craft licensing, fishing permits, landing of catches, limits to size of caught fish, limits to size of mesh and fishing nets, prohibition of the use of explosives in addition to offenses and penalties for non- compliance. The ordinance was revised and amended twice into "Marine Fisheries Act" in the years 1960 and 1975. However, the Marine Fisheries Ordinance and Marine Fisheries Act were somewhat limited and did not cover all fisheries resource regulatory aspects. The 1975 bylaw included a specific article addressing pollution and another item banning the collection of corals and organizing the catch of ornamental fish. The Regulation of 1975, which succeeded the Marine Fisheries Ordinance of 1937, aimed to regulate fishing and the use of marine resources in the territorial waters of the Sudan. There is an urgent need for a comprehensive review of acts and bylaws in affect to streamline the process of fishing management and to allow for a more dynamic approach to sustainable utilization of the resource.

Other relevant developed legislation includes the following:

- Marine Act – 1961
- Territorial Water and Continental Shelf Act – 1970
- General Regulation & Control of Merchant Shipping – 1971
- Petroleum Act – 1998
- Environment Protection Act – 2001

The above mentioned acts and regulations are not specific to fisheries resources management but deals with the marine resources in general which in turn impact the fisheries directly or indirectly.

Sudan is also an active member of the following organizations:

- Committee on Inland Fisheries and Aquaculture of Africa (CIFAA)
- Indian Ocean Tuna Commission (IOTC)
- Regional Organization for the Conservation of the Environment of the Red Sea and Gulf of Aden (PERSGA)

Limited Research Capacity

There are few research entities engaged or concerned with research relating to the Red Sea. The Marine Research Centre is the main body in charge of marine research. In addition, there are also two specialized research stations that work under the Red Sea Fisheries Administration, namely the Fisheries Research station which is mainly concerned with marine fisheries research, and the Aquaculture Research Station which focuses primarily on fish, shrimp and pearl-cultivation research. Research in fisheries is also undertaken by some universities, especially those that are marine and fisheries research-oriented such as the Red Sea University, Khartoum North University, Sudan University for Science and Technology and at some other higher education providers within the country. They also contribute to the development of the marine and fisheries sector by offering specialized programs in Marine Sciences, Fisheries Science and Environmental Management. Graduates of these programs contribute to the work force in the marine and fisheries sectors.

One of the main constraints in developing research capacity are the limited allocated funds for research. Other limitations include poor infrastructure, and limited opportunities for training abroad and capacity building programs in general. Currently, the research stations and national researchers are mainly involved in the ongoing NGOs, international agents' projects and bilateral governmental projects. The exercise is providing mutual benefits for both parties.

Natural and Environmental Challenges

Most of the fishing activities that occur in the Sudanese Red Sea area are inshore, and are therefore concentrated within the reef area. Illegal fishing methods, and over fishing in such areas have direct negative impacts on the coral reefs and other biological communities. The expansion in the marine sector and coastal physical infrastructure have led to some environmental problems such as pollution and habitat destruction (Abdellatif, 1993; CIDA, 1995; Abu Gideiri, 1999).

Natural constraints include bad weather conditions (monsoons, excessive heat and storms) which directly affect the fishers and fishing trips. One of the recent developments is the introduction of a weather forecast service for fishermen to help them plan their fishing trips and activities (FAO, 2016).

Environmental legislation has been introduced in the year in 2001. It sets the legal framework for environmental protection including the marine environment. Sudan is committed to environmental conservation and sustainable management of the resources and therefore is an active member of some regional and international agreements that are listed below:

- Convention on the Protection of the World Cultural and Natural Heritage -1974
- The United Nations Convention of the Law of the Sea – December, 1982
- The Regional Convention for the Conservation of the Environment of the Red Sea and Gulf of Aden – 1984
- The United Nations Convention on Biological Diversity – 1992
- Convention on International Trade in Endangered Species of Wild Fauna and Flora – 1982

SUPPORT BY INTERNATIONAL AGENTS AND ORGANIZATIONS

The following provides examples of international organizations' contribution to the development of the fisheries sector. Bilateral support has been provided by many international agents and organizations. These include, but are not limited to the following examples:

Historically, the Food and Agriculture Organizations (FAO) had played a vital role in support of the country, especially in terms of the development of the agriculture and fisheries sectors as a means of food security. Its role has been instrumental in the progress and recent development of those two sectors.

The British Overseas Development Administration (ODA) executed a project for developing the Red Sea artisanal fisheries through the building and mechanization of boats, and the provision of workshop, maintenance, storage and transport facilities

The Canadian International Development Research Centre (CIDRC) has provided great support and assistance in the field of aquaculture, to increase fish production, and to develop suitable technologies for oyster culture.

OPEC/UNDP have concluded a development project in the Red Sea, mainly through cooperatives support, fishing gear improvement and assistance in the fields of storage and transport of fish in the northern parts of the Sudanese Red Sea.

The United Nations Development Program is providing technical support and assistance for the assessment of fisheries' potential.

UNIDO with the support of the Canadian Government has also been working in a multi -faceted program for the development of the Sudanese Red Sea fisheries sector.

The Committee of the Indian Ocean Tuna Commission is also working with the Red Sea Research Centre with a main focus on stock assessment and management of Sharks fishery (FAO, 2017).

The Norwegian Government has established a fully funded project mapping the living marine resources in the Sudanese Red Sea coast.

A notable challenge that needs to be highlighted is an unintended consequence of international organizations' support efforts. Despite the great progress that is often achieved when international organizations are involved in supporting the fisheries sector in Sudan, these efforts have often gone to waste upon the expiration of the organizations' operation period. It has been continuously observed that once the support period comes to an end, the projects often fail shortly thereafter. Going forward, it is vital for the local policy makers and fisheries sector administrators to be aware that unless adequate training and hand-over support are provided, projects introduced by external organizations will fail to deliver the intended outcomes.

Current Standing and Future Directions

In order to assess and evaluate the fisheries' current status and future direction, it is imperative to evaluate the government's strategies. The Quarter Century Strategy of the Sudan (2002-2027) is a comprehensive long term national strategy that sets forth strategic goals and objectives as the guiding principles for developing and strengthening the various sectors including the fisheries. These goals are listed below:

- An enhanced role for fisheries in poverty alleviation, food security, human health and environment.
- Adopting scientific research and technology advancement as vehicles for increasing productivity and efficiency.
- Rational utilization, conservation and development of aquatic and fisheries resources through sustainable production management, restocking of depleted fish stocks and pollution control.
- Strengthening economic infrastructure and promoting privatization.
- Strengthening public and private sector institutional setups.
- Securing participation of fisheries sector beneficiaries in the management and development processes.
- Developing and strengthening the competitiveness of fisheries products through improvements in marketing channels, quality control and safety.
- Promoting sustainable development.
- Strengthening and developing information resources and databases.
- Strengthening regional and international cooperation, including agreements, exchange of experiences, joint programs and scientific forums.
- Institutional and legislative reforms.
- Strengthening coordination mechanisms between the public and private sectors at the central and state levels within the country.
- Establishing and developing fisher and producer organizations.
- Promoting fish producers and fisheries investors through stimulating easy-term credit systems.

The main lines of emphasis are highlighted below:

- Ensuring food security and curbing poverty.
- Promoting sustainable development through sound utilization and conservation measures of the fisheries' resources.
- Strengthening databases and linkages, capacity building and technology transfer.
- Empowering institutions and their support capabilities.
- Promoting Research and Community services.
- Promoting trade and marketing strategies.

CONCLUSION

The Quarter Century Strategy of the Sudan (2002-2027) reflects the country's commitment to the sustainable development of its resources with a main focus on ensuring food security and curbing poverty. Food security, poverty alleviation, and improving the socio-economic status are the most pressing issues that require immediate attention and prompt actions. The strategy also indicates that policy makers are aware of the existing and anticipated challenges. However, in order to ensure the success of the aforementioned strategy, it is essential that clear and coherent policies are created and publicized. Such policies would allow for proactive and logical actions to be taken, which would align with, and support the fulfillment of the key strategic goals.

Nevertheless, it follows from the above review that while there has been notable progress and achievement during the last few years, some challenges remain. These challenges can be addressed by taking the following actions:

- Developing a robust system for data collection, recoding and documentation.

- Setting up programs and action plans in line with the strategic objectives with adequate provision for implementation and follow up mechanisms.
- Reviewing and updating the fisheries legislation and the closely related bylaws to be more responsive to recent developments and evolvement of management practices and control measures.
- Adopting a holistic approach for ongoing and future projects to avoid the shortcomings of previous projects that have not been successful due to their failure in addressing the root causes.
- Diversify market and marketing services to introduce more competitive forces through new channels of distribution.
- Modernizing the fisheries sector by introducing better methods of harvesting, processing, and introducing quality standards.
- Restructuring the management system and ensuring adequate horizontal and vertical collaboration within areas of competence.
- Strengthening governmental institutions through training and capacity building programs.
- Ensuring that effective management policies are in place, which is critical to the development of the fisheries resources, optimal utilization, and sound economic returns.
- Enforcing the FAO code for responsible fisheries.
- Facilitating communication between the relevant stakeholders.

REFERENCES

Abdellatif, E.M. (1993). *Factors Threatening the Marine and Coastal Environment of the Red Sea in Sudan, A Call for Attention*. Environment Conservation Society. Khartoum, Sudan.

Abdulla. A.N.E. (2013). *Sudan National Report to the Scientific Committee of the Indian Ocean Tuna Commission*. Fisheries Research Centre. Port Sudan, Sudan.

Abeywickrama, K. (2009). *Market Assessment & Recommendations: Recovery of Coastal Livelihoods in the Red Sea State of Sudan through the Modernization of Artisanal Fisheries and the Creation of New Market Opportunities*. UNIDO Project TF/SUD/09/002-1154.

Abu Gideiri, Y.B. (1999). *Biodiversity in the Red Sea*. National Biodiversity Strategy and Action Plan (NBSAP), Khartoum, Sudan.

Altayeb, M. (2001). *Biodiversity of Echinoderms at Abu Hashish Fringing Reef*, Sudanese Red Sea Coast. M.Sc. Thesis, Institute of Environmental Studies, University of Khartoum, Khartoum. Sudan.

CIDA. (1995). *Preliminary Environmental Assessment: Sudan Integrated Fisheries Project*. Canadian International Development Agency. Halifax, Canada. Pp 50.

Crossland, C. (1957). The Cultivation of the Mother of Pearl Oyster in the Red Sea. *Austr. J. Mar.* Freshwater Res.

Country Brief. (2014). [online]. Available at: http://cbs.gov.sd/files/pop-proj-by-states137.pdf.

FAO. (2008). *Country Profile Sudan*. [online] Available at: ftp://ftp.fao.org/fi/document/fcp/en/fi_cp_sd.pdf.

FAO. (2016). *Fisheries and Aquaculture Information and Statistics Service*. [online] Available at: http://faostat.fao.org/site/610/default.aspx#anchor.

FAO. (1984). *Yearbook of Fishery Statistics. Catches and Landings*. Food and Agriculture Organization of the United Nations. Rome, Italy. Vol 54.

FAO (2o17): *Sudan Inland and Marine Fisheries*. Available at: http://http:// www.fao.org/fi/FCP/FICP_SDN_E.ASP.

Farah, O. M. (2000). *Sudanese Marine Fisheries Statistical Data*. Prepared for PERSGA/SAP First Working Group Meeting on Standardization of Living Marine Resources Statistical Data (Fisheries). Fisheries Training Institute. Aden, Yemen.

Khalil, A.S. and Krupp, F. (1994). "Fishes in the Mangrove Ecosystem" In: *Comparative Ecological Analysis of Biota and Habitats in Littoral and Shallow Sublittoral Waters of the Sudanese Red Sea.*

Forschungsinstitut Senckenber. Frankfurt, Germany, and Faculty of Marine Science and Fisheries. Port Sudan, Sudan.

Osman, M.S. (2004). Country Review: Sudan. *Review of the State of World Marine Capture Fisheries Management*. FAO. [Online] Available at: http://www.fao.org/docrep/009/a0477e/a0477e0t.htm

Randall, J.E. (1992). *Red Sea Fishes*. Immel Publishing Limited. London, UK.

Sanders, M.J.; Morgan, G.R. (1989) *Review of the fisheries resources of the Red Sea and Gulf of Aden*. FAO Fisheries Technical Paper. No. 304. Rome.

Sheikeldin M.E. (2014). *Biometric Relationships of the Mother of Pearl Oyster* (*Pinctada margaritifera*). Red Sea University. Online. Available at: http://www/researchgate.net/publication.

UNIDO (2009). Inception Report *Recovery of Coastal Livelihoods in the Red Sea State of Sudan*. Red Sea State, Sudan.

United Nations (2017). *World Population Prospects - Population Division - United Nations*. [online] Available at: https://esa.un.org/unpd/wpp/DataQuery/

INDEX

D

E

F

G

H

I

L

M

N

O

P

Q

R

S

T

V

W

Y

Z